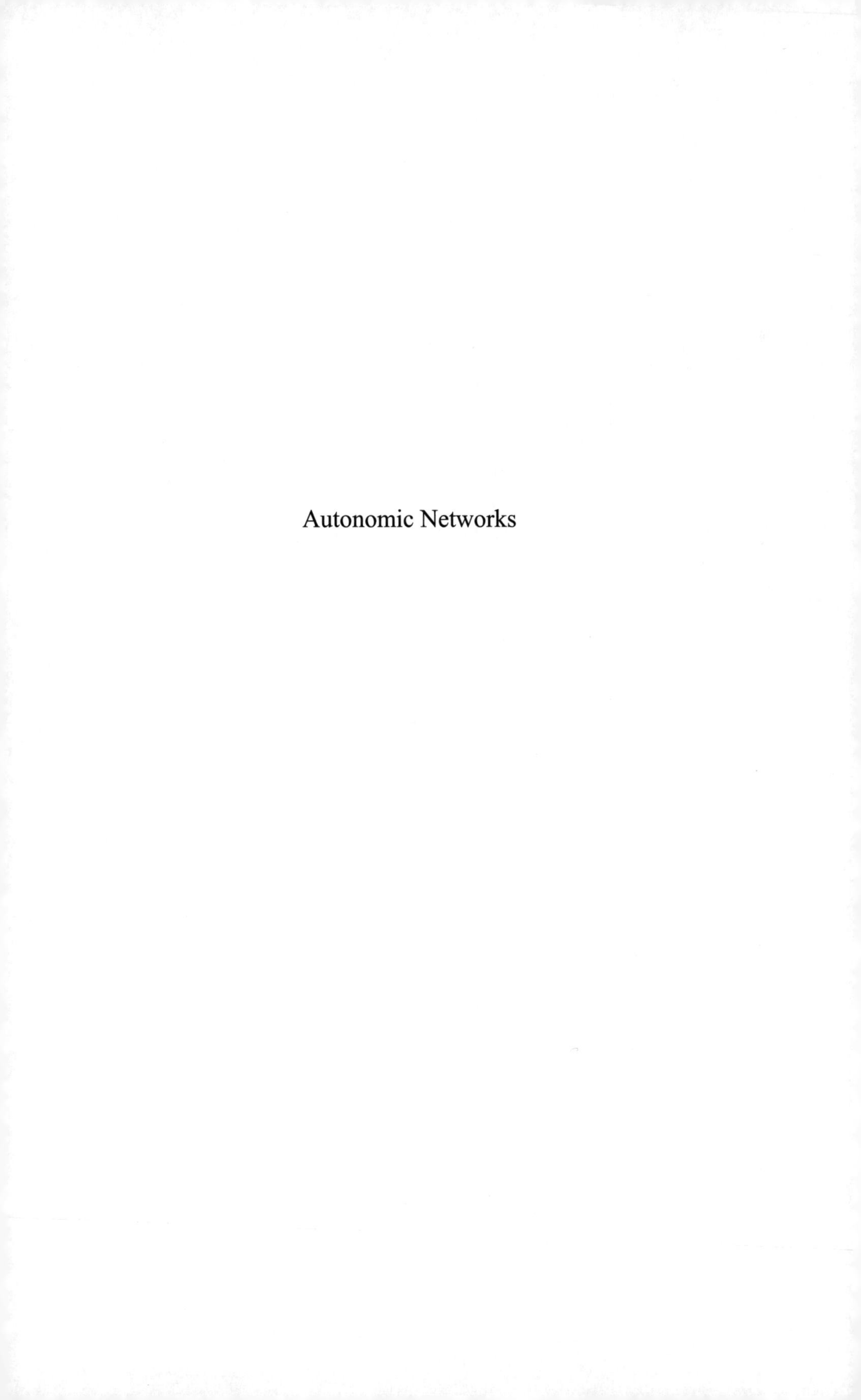

Autonomic Networks

Autonomic Networks

Edited by
Dominique Gaïti

Part of this book adapted from "Intelligence dans les réseaux" published in France in 2005 by Hermes science/Lavoisier
First published in Great Britain and the United States in 2008 by ISTE Ltd and John Wiley & Sons, Inc.

ISTE Ltd
6 Fitzroy Square
London W1T 5DX
UK

www.iste.co.uk

John Wiley & Sons, Inc.
111 River Street
Hoboken, NJ 07030
USA

www.wiley.com

Library of Congress Cataloging-in-Publication Data

Autonomic networks / edited by Dominique Gaïti.
p. cm.
Includes index.
ISBN 978-1-84821-002-8
1. Telecommunication--Computer programs. 2. Computer networks. 3. Distributed artificial intelligence. I. Gaïti, Dominique.
TK5102.5.N5238511 2008
621.382'1--dc22

2007018294

British Library Cataloguing-in-Publication Data
A CIP record for this book is available from the British Library
ISBN: 978-1-84821-002-8

Printed and bound in Great Britain by Antony Rowe Ltd, Chippenham, Wiltshire.

Table of Contents

Introduction

The Internet network is a difficult network to control because of its initial design where only one service class exists: the *best effort* service, which requires the network to do the best it can for all the users. It is an excellent solution for keeping the connection costs low since control is done at the terminal machine, PC or other mobile terminal level.

On the other hand, for land or mobile users, the Internet network is not always the most convenient solution. In fact, each network application must be able to move data packets with a quality of service which guarantees an adequate performance for the application. In order to achieve this, innovations in the last few years have made routers and network protocols more complicated by introducing, for example, priorities between the different moving flows. The main disadvantage is the increase in network complexity thus making it harder to manage and control. Furthermore, the optimization of network resources based on user requirements is not well managed, thus leading to a waste of network resources which can become costly.

After the propositions failed to introduce quality of service to users at a reasonable cost, network research specialists have focused on new opportunities in order to go back to end control for quality of service and to optimize network resources. In addition, newly implemented technologies are also designed to guarantee communication security and user mobility management.

At this point, these new technologies mainly come from the artificial intelligence world. This research leads to the emergence of a whole environment of intelligent software entities within an architecture that is adaptable to a particular context and environment. We use the term "intelligence" to recall the adaptation capacity shown

---

Introduction written by Dominique GAÏTI.

by an entity. An intelligent entity is then an entity which is able to adapt to a context, to a particular environment and/or to events that may never have been encountered.

This research is an extreme innovation at present but it is difficult to implement because it calls for competences coming from both the networking and artificial intelligence worlds. The pairing of these two domains will lead to:

– a dynamic and intelligent control of local equipment;

– a global control of the network in a cooperative way;

– a more autonomous network management;

– a better guarantee of end-to-end quality of service.

The technology used, on which these new adaptable and even autonomous networks are based, mainly comes from new research on intelligent agents and multi-agent systems. This technology is tested today in widely diverse areas: computer-aided design, social behavioral simulation, intelligent management of distributed systems, image recognition, etc. This research has demonstrated the efficiency, reliability and robustness of multi-agent systems for the dynamic control of complex and distributed systems. IP network infrastructures now appear as a natural application domain for software agents in general and particularly for cooperative software agents. Despite the hesitations from certain players in the field, this technology is slowly finding its place in telecommunication networks.

Multi-agent technology is an innovation of the last 10 years emerging from several research fields: symbolic artificial intelligence, the theory of control and distributed artificial intelligence. The quick development of this technology is due to the necessity for new solutions and new mechanisms to resolve complex problems. This technology is slowly becoming key as computer systems increasingly become distributed, interconnected and open. In such environments, the capacity of software agents to plan their actions and goals, to cooperate and negotiate autonomously with others, and their capacity to respond in a flexible and intelligent way to dynamic and unpredictable situations will help to reach a significant improvement in the quality and performance of software systems.

The final challenge of all these studies is to make the network completely autonomous.

This book is designed to show the different aspects that the research conducted has uncovered until now in order to reach this ultimate goal of autonomy. The different integration studies conducted have mainly focused on the management and control aspects in the IP network field, whether wireless or not.

Chapter 1 gives an overview of network management and monitoring aspects and of artificial intelligence techniques and control which may be considered in this context. The evolution from expert systems to mobile agents via multi-agent systems is presented briefly.

The first form of intelligence integrated to the network, although not directly based on artificial intelligence concepts, is addressed in Chapter 2 through the notion of active networks. The active network is used here to reach an active control of the quality of service in an IP environment and thus ensure an adaptive management.

Chapter 3 explains in more detail the concepts of Chapter 1 on agent aspects and their adequacy in dealing with the problem discussed after putting an emphasis on IP management problems. The possible applications for artificial intelligence-network pairing are also addressed: simulation, quality of service, continuity of services, congestion control, monitoring, topology maps or routing.

Chapter 4 presents an advanced approach in terms of wired or wireless network management called policy-based management. The use of intelligent agents in this context is discussed. A new application is highlighted, i.e. the service contract negotiation between a client (mobile or not) and a service provider, and between two service providers.

From Chapter 5 onwards, the agent concept based on artificial intelligence is enhanced by multi-agent systems and associated platforms. Their evaluations, which are often considered as the weak point of these systems, are addressed.

Chapters 6, 7, 8 and 9 together form a coherent group of studies conducted in the LIP6 laboratory of UPMC (University of Paris 6) and at LM2S of UTT (University of Technology, Troyes) among others. Advanced research on artificial intelligence has been conducted particularly in behavioral management (behavior based on human behavior) from DiffServ network elements (nodes, routers, etc.). This research has been able to prove that the use of intelligent agents in the analysis and management of DiffServ networks enables us to understand failures in a more precise way and thus to be able to significantly optimize these networks (Chapters 6 and 7).

Research projects have mainly focused on the multi-agent simulation of IP networks and have led to the creation of a simulator which makes it possible to analyze networks that support DiffServ QoS and intelligent agents. This piece of software can simulate any type of model (network or not) and, although it is still new, it has serious advantages (changing parameters during simulation, portability, reliability) which ensure an ongoing research as well as its use in the industrial field (Chapters 8 and 9).

The advantage of this technology is the consideration of active elements leading to the operation of networks. All the parameters can dynamically change in time and space. These new parameters have never been considered in a modeling, simulation and control context before.

Chapter 10 addresses a particularly crucial problem that is the management of mobility in the 3rd and 4th generation. Here again the use of agents in this context brings significant added value, particularly in the VHE (virtual home environment) domain.

Chapter 11 completes the previous chapter by integrating the concept of learning within the problem of dynamic allocation of radio resources, which are particularly expensive and therefore are imperative to optimize. Learning is without a doubt the ultimate intelligence step since it is this step that will enable the systems concerned to adapt to an environment and to be able to attain the concept of autonomy.

Finally, Chapter 12 somewhat integrates the different aspects discussed throughout this book and presents a possible example of architecture which combines concepts of management and control, of policy-based management, of active network and of mobile code in a DiffServ environment thus leading to an automation of the different control points.

All the studies presented in this book mark a step in the research conducted on embedded intelligence in a telecommunication network environment. The ultimate step will lead to an autonomous network, which is considered as one of the most important undertakings for the next 10 years in terms of telecommunications. Even though the main activity is currently in research, we should still mention the emergence of the first start-up (Ginkgo Networks) devoted to the autonomy of agent networks at the end of December 2004.

Chapter 1

Artificial Intelligence and Monitoring of Telecommunications Networks

1.1. Introduction

During the last decade, telecommunications have experienced several important developments: the gradual disappearance of monopolies in Europe, the explosion of mobile and Internet markets, the deployment of broadband networks (ATM, ADSL) and intelligent networks to name but a few.

The elements that make up these networks are not exempt from failures and they can experience several malfunctions. Therefore, it is no wonder that monitoring and improving the networks to prevent technical hazards are so important to operators. In order to achieve this, they establish a series of deductions enabling them to diagnose the nature of the unusual behavior as opposed to the simple detection of anomalies appearing on the network in a passive manner. This form of active detection, paired with a diagnosis of failure, is what we mean by monitoring. In order to limit expenses and have a complete view of the network, telecommunications operators centralize their monitoring tasks. The supervisors in a center are responsible for making sure the global operation of their network is adequate.

However, this role is complex because a telecommunications network is made up of a large number and variety of components, which are spread over vast territories, transmitting messages to the monitoring center in order to report changes in

---

Chapter written by Hassine MOUNGLA.

behavior (halt, anomaly, problem, etc.). Numerous messages can arrive at the monitoring center, which must process them in real-time, all the while maintaining an adequate quality of the services provided. Network monitoring tools have been designed for these reasons.

Among these tools, several implement techniques which are based on artificial intelligence (AI): rule-based systems (expert systems), communicating agent-based architecture, model-based methods, automatic learning, etc. This chapter presents AI tools which are specifically designed to assist in monitoring telecommunications networks. These tools must fulfill two main requirements: efficiency and adaptability. Efficiency is essential to process in real-time the mass of information coming into the monitoring center. Adaptability is necessary to monitor the structural and physical evolution of the networks.

This chapter is divided into two parts. The first part is dedicated to the issues and we will address monitoring problems and goals. The second part presents the main AI techniques that apply to this field. In conclusion, we will present an assessment of AI's contribution to telecommunications network monitoring.

1.2. Network management goals

A general overview may show a network user expecting a quick and reliable connection, a manager wanting to easily configure and give access to resources, and an operator wishing for a low usage cost. However, a more detailed view, which is proposed by [TER 92] and reported in [STA 96], establishes the network management requirements as follows:

– complexity control: continuous growth of the number of components, users, interfaces, protocols and network providers threatens the management with the loss of control of everything that is linked to the network and of the way the network resources are used;

– improvement of services: users expect an improved service such as quality of service and calculation resource;

– control of the company's capital strategies: networks and distributed computing resources are vital resources for most organizations. Without efficient control, these resources do not provide the performance needed for a company;

– balance requirements: the information and IT resources of a company must provide support for a wide range of users with specific conditions on performance, availability and security. The network manager must assign and control resources to balance these various needs;

– decrease of breakdown time: when the network resources of a company increase, the minimum availability conditions must be close to 100%;

– understanding control costs: the use of resources must be monitored and controlled in order to satisfy the user's needs at a reasonable cost.

Due to their heterogenity, networks are difficult and complex systems to manage. Management systems must handle a large number of devices, resources and protocols over extended areas. That is why the ISO/TC 97/SC 21/WG4 workgroup was put together within the ISO (International Organization for Standardization) in the 1980s [PRA 95]. In 1987, the IETF (Internet Engineering Task Force) made network management propositions for the management of the Internet. The protocols SNMP and MIB (*management information base*) were normalized and quickly implemented on all the network's devices, and have become the norm. However, contrary to ISO, the IETF does not provide a management architecture, which means its standards are difficult to implement, particularly in WAN (*wide area network*) management.

1.3. Monitoring needs of telecommunications networks

In this section, we will present the needs of users and service providers for a successful deployment of telecommunications services such as those described in the studies of Semoude [SEM 00]:

– the need for a quick introduction of services in network infrastructures. For this, reusing components is a significant criterion. Indeed, the abundance of object models and the presence of library/packages of useful and easy to use function/object/components facilitates the creation of new services and their deployment in network architectures;

– the need for simple tools for the deployment of services with QoS constraints and cost minimization. In fact, the distribution of service elements must be optimized in order to dynamically compensate for the network's failures;

– the need for personalized QoS which was made necessary by the evolution from a static SLA to a dynamic or proactive SLA in order to adapt to users' requirements;

– the need for service management architecture. The management of services is defined as a group of principles and mechanisms that ensure monitoring, analysis, coordination, control and planning of the operations for the different entities which are part of the services (resources, software components, etc.). The service management is separated into managing and managed components which are the service elements deployed over different network nodes. The management architecture must provide:

- a set of specific rules and instructions for designing the system that will provide the services,

- the possibility of identifying terminal equipment,

- quality of service control,

- invoicing, including invoicing in a heterogenous context,

- protection for the entities providing the service and their interactions,

- openness to support cooperation, i.e. an architecture which is able to interoperate with other architectures and other systems through interfaces.

Finally, what we believe is the most important characteristic: flexibility of telecommunications network monitoring and management architectures. This flexibility is measured by the degree of adaptation of these architectures in unknown situations.

Table 1.1 lists a few telecommunications architectures that we will not attempt to explain in detail.

Type of architecture	Service oriented	Structuring element for the services	Management network	Component oriented
Architectures	IN UPT UMTS	OSA TINA OSAM	TMN	ODP OMG/OSM

Table 1.1. *Types of telecommunications architectures*

Certain architectures take services into consideration (intelligent networks [UIT Q1201, UIT Q1202], UPT, UMTS, etc.), whereas others are more general (OSA [CAS 94, CAS 95, CAI 94], TINA [TIN], OSAM [DOL AC036], etc.) and propose structuring concepts for the current needs (multimedia, voice/data, etc.). The OMG/OSM [OMG 00] and the UIT-T/ODP [ZNA 97] introduce the possibility of handling the objects, knowing that TMN [UIT 96] is an architecture which is more specific to network management.

Traditionally, the telecommunications market was dominated by large providers, with proprietary hardware and software systems. Any change offered by a network operator on telecommunications services required negotiations with several providers in order for them to adapt their hardware and software as needed. This was a long and expensive process that needed to be cut back.

Intelligent networks (IN) constituted the first initiative for defining an open architecture in terms of reusable software entities. For new services, IN is based on the conceptual INCM (*intelligent network conceptual model*) model. This model separates the logical and physical aspects and encourages the creation of services. It is only in the last 10 years that the emphasis has been placed on the cooperative management of network and services leading to telecommunications management network norms (TMN) which are presented in section 1.4. In this context management provides functions such as FCAPS (fault, configuration, accounting, performance and security), SMF and changes in routing algorithms as well as user management (subscribers) [UIT 96]. The adoption of TMN norms satisfies the management concept needs of network elements, networks and services hierarchy. TMN encompasses the concepts of management domains where one domain can be considered as a specific network technology. Propositions have been made by international organizations to coordinate IN and TMN approaches either by interoperability between the two, or by the development of an extension of TMN to cover IN functions (for example, the control of the service can be treated as a real-time service control).

The object-oriented approach and the distributed process are technologies that have enabled significant advances in network management. The norms for open distributed process (ODP) (recommendation from ITU-T, X.900 series) provide a structural framework. ODP identifies a model of network infrastructure with standard services and components supporting the distributed process. ODP's openness has become the concept which distinguishes teleservices, bearer services and is one of the main goals of service architectures. The notion of service architectures was introduced by BELL within the INA project (1989-1991) and by the RACE-ROSA European research project (1989-1992), and was then carried on by a certain number of other projects, in particular TINA-C (1993-1997) [SEB 98].

1.4. The telecommunications management network (TMN)

Network operators are responsible for maintaining a variety of networks for applications such as public telephony, slow and high speed data transmission, mobile radiophony, message switching, etc. In each case, they must operate and maintain switching and transmission equipment. At the same time, it is also important to continuously improve the functions which maintain the numerous advanced services that users and service providers are demanding [ITU 96]. TMN offers a global solution for maintenance, administration and operation of networks from any type of technology.

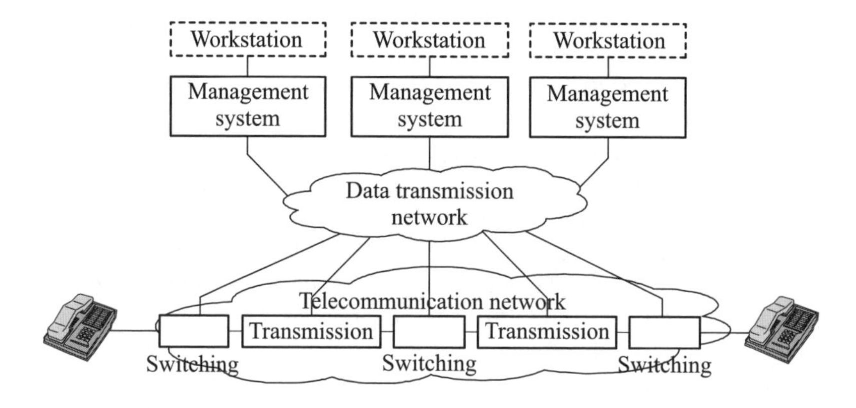

Figure 1.1. *Relation between TMN and telecommunications network [SAM 01]*

The telecommunication management application is made up of agents spread over the monitored network. These agents communicate in order to coordinate their actions. These communications can only happen when there is a network linking the agents. This network is a fundamental management concept and its principle is presented in the M3100 recommendation of the ITU-T. A TMN provides a network structure interconnecting operating systems and standardized telecommunications equipment.

1.4.1. *TMN management functions*

In order to carry out its activity, the TMN relies on a certain number of functions which can be grouped in five functional areas, or SMFAs (*specific management functional areas*), and defined by OSI control in the M3100 recommendation of ITU-T:

– fault management;

– configuration management;

– accounting management;

– performance management;

– security management.

These serve as a support for the analysis of network support applications such as taxation management, traffic management and even network planning.

1.4.2. *TMN architecture*

There are three basic aspects within the TMN architecture:

– functional architecture: this describes the adequate distribution of functional elements within TMN. This is adequate in the sense that it will create function blocks from which it is possible to implement a TMN no matter how complex it is [ITU 96];

– TMN information architecture: this is based on an object-oriented approach. The management principles of OSI systems communicate with TMN principles and, when necessary, they extend to adapt to the TMN environment;

– physical architecture: this describes the possible interfaces and gives examples of the physical components within TMN.

1.5. Control in telecommunications networks

In order to satisfy the increasing application needs while ensuring an efficient use of available resources, telecommunications networks must be adequately *controlled*. This control can take multiple forms and take place at different levels. The diverse forms of control do not all use the same type of information and do not all work on the same timescale. Finally, the different control types are not systematically in all the networks; their use is often specific to the information transport mode.

First, we have *admission control*, whose role is to accept or refuse the connections for services functioning in connected mode (in theory a network in stand-alone mode accepts all the traffic offered and attempts a fair share of its resources). This decision depends on the characteristics of the active connections and the state of the network. Specifically, control devices base their decision on statistical information which supposedly characterizes the network and the source during the time of a new communication (this task is not executed based on the instant state of the system). One of the main problems in implementing this type of control comes from the fact that the effect of the decision cannot be reversed and that we have to consider, for example, possible increases in traffic that will compete with that of the communication that we have just accepted.

The network then manages or controls its operation at the *routing* level by selecting the route that the information will follow. This control is based on several criteria and considers the global state of the system; the connection mode also affects the routing algorithm. Routing can be static or dynamic. The static routes are fixed in the system and manually or very slowly modified. When algorithms evaluate the routes between two sites in a permanent manner, routing is then called dynamic. For stand-alone networks such as IP networks, dynamic routing can affect the route of each transmitted packet. Two consecutive packets cannot follow the same route because of a change in the tables (and arrive at destination in the reverse order from that of their transmission). In the case of connected networks, the circuits are rarely modified during their lifetime (except in certain cases of telephony where rerouting is done). This does not prevent having a dynamic routing algorithm in these networks. When the decision to modify routing is taken in order to avoid network saturation (usually in the case of packet switching), it is called *congestion control*. The term congestion control is also used for the mechanisms used to restrict access to an overloaded resource in the network. This type of control can be seen as one of the aspects of the more general network administration concept.

Networks operating in datagram mode or by virtual circuits can decide to limit the number of information units which are sent at each moment in order to avoid the appearance of congestion points. This is called *flow control*. These decisions are generally made in real-time based on instant state indicators. The first example of this type of mechanism is the control by *sliding window* at link level or end-to-end. A technique which is similar and particularly important for ATM networks is *traffic shaping*, where we want to modify certain traffic characteristics exiting a source so that it will re-enter the network by verifying given properties.

We must not mistake flow control with congestion control: the more general flow control is in place to ensure congestion control. The goal of congestion control is to ensure that the network can transport the traffic offered based on specifications. Flow control is always linked to a source and a receiver. We can classify flow control techniques as open loop techniques, as in *traffic shaping*, and closed loop techniques, for example, sliding windows. We should mention that when we talk of open loop flow control, we integrate the previously mentioned admission control techniques.

Finally, the capacity sharing problem of a node must be considered. The node must decide what fraction of the bandwidth and storage resources must be allocated. Furthermore, the communication must adapt in real-time to the state evolution of the system. This problem is generally associated with a connected transmission but not exclusively (see the RSVP (resource reservation protocol) in the Internet network's projected evolutions).

We can observe that particular problems can appear when we connect networks of different natures, particularly if they operate in distinct modes, for example, routing of IP packets (datagram mode) over an ATM transport network (virtual circuit connected mode). Sophisticated control mechanisms must be implemented to take this support heterogenity into account and in particular to ensure the quality of service (QoS) expected by users.

1.6. Some AI techniques for monitoring telecommunications networks

Monitoring a network consists of following its evolution as alarms arrive, displaying the failures and possibly suggesting certain actions to the operator.

Telecommunications network monitoring has until now used expert systems (for alarm detection) (see [EUR 91]) among AI techniques. In section 1.6.1, such a system is presented. However, the drawbacks of expert systems (ambiguous explanations, low genericity, expertise cost and delay, difficult update, etc.) have led to the development of other monitoring methods, in particular model-based diagnosis techniques ([DKL 87, REI 87, CON 91, BRU 96], or [AZA 93, DBE 95, PFA 93, RIE 93a]). In section 1.6.2, we present the studies from CNET relying on these basic model techniques to ensure monitoring.

1.6.1. *Chronos: an expert system generator for monitoring telecommunications networks*

Chronos [EUR 91] is a real-time expert system generator. Initially a first IROISE prototype was developed in LISP at CNET around 1989. Toward 1992, a new version, called Chronos, was developed. A third version was designed in NM-expert for the evaluation of the NM-expert language in the case of a real problem.

An expert system written in Chronos is made up of a set of rules. A typical rule contains an assumption part and a conclusion part. The assumption part describes a series of alarms, indicating for each one its nature, source and possibly its date of arrival and the delays between these dates, or again an indication of a minimum number of alarms of a given type which must be received within a given delay. The conclusion part generally indicates the undesirable event which occurred in the network and which is presumed responsible for the transmission of alarms. It can also state the actions that the supervisor should take (commands to send). A typical example of rules is given in Figure 1.2. For legibility reasons, the rule is not shown in Chronos but in an informal language [MAY 99].

When:

We receive an alarm X for a technical center at time T1

and:

we receive an alarm Y for this technical center at time T2

where T2 > T1

and:

during the time period [T2, T2 + 30 s], we receive more than 3 alarms of a certain type with a switch in this technical center

then:

display "there has been a halt of TC from time T1 to time T2"

Figure 1.2. *Expert system rule*

The main known cause is technical center (TC) halt. The acknowledgement is done from a sequence of alarms and includes temporal data [MAY 99, ROZ 97]; we can then conclude that the technical center has stopped operating.

The quality of an expert system mainly depends on the quality of expertise used to design it. A second disadvantage comes from a lack of genericity: the expert system depends on the network, its configuration, its components, their operation and delays. The modification of one of its elements requires the development of a new expertise. One last drawback is the lack of legibility of the conclusions drawn by the system. For example, from the rule given in Figure 1.2, it is impossible to understand why we conclude that the technical center has stopped operating.

1.6.2. *Monitoring with model-based techniques*

Another network monitoring method is proposed [UNG 93]. It is based on model-based techniques from two distinct characters:

– *consistency-based* character [REI 87, DK 87] where the goal is to find the state of the system matching the observations; this approach is based on a model of correct system operation;

– *abductive* character [CON 91a, CON 91b] where the goal is to explain the observations. This approach is based on the model of system malfunctions.

We will not go into a deep analysis of model-based diagnosis (for an overview of the current techniques see [HAM 92]). The advantages and disadvantages of systems using deep understanding are presented in [UNG 93]. The main advantages are [ROZ 97, MAY 99]:

– genericity: in order to apply the diagnosis on different systems, we must have a model of these systems;

– upgradeability: the update of the model following the system's upgrade;

– ease of development;

– reuse of diagnosis;

– the quality of the explanations.

The main disadvantages are:

– the calculation time required to carry out the diagnosis, which is cumbersome for a real-time operation;

– the construction of the model, which in the case of real problems becomes a difficult task.

We should mention the GASPAR (gestion d'alarmes par simulation de pannes de réseaux (management of alarms by network failure simulation)) project whose goal was to implement a generic and efficient monitoring from previously presented techniques [ROZ 97, MAY 99].

Other projects have also been developed with the help of model-based techniques. We can mention the MAGDA [MAG 03] and MAGDA2 (modélisation et apprentissage pour une gestion distribuée des alarmes de bout en bout (modeling and learning for a distribution management of end-to-end alarms)) projects.

MAGDA was developed from a business need: the correlation between alarms and diagnosis. The main results of this project are:

– development and prototyping of a completely new approach in the field of alarm and diagnosis correlation. A new *distributed* diagnosis technology based on modeling resembling the management system was developed;

– knowledge acquired in crash and alarm modeling in SDH networks. In addition to modeling itself, the development methodology of this modeling constitutes a transportable benefit;

– transfer of a new event correlation technology, i.e. *chronicle recognition*, which was developed by the FTR&D laboratory and integrated by ILOG in its "Rules for Telecom" offer. The preindustrial prototype nearing completion in MAGDA will enable *JRules* to provide an extended capacity of expression through the notion of chronicle (partial sequence of dated events). The integration of the *chronicles* technology in the standard version of JRules will be decided depending on the studies conducted with the *beta* prototype;

– extension of the *Almap-IF* platform from Alcatel to CORBA communication (in addition to the communication by Q3 format used in network management). This has enabled the development of an architecture prototype where the diagnosis module can be integrated, with *Rules* at the front for filtering purposes.

1.6.3. *Agent technology*

The introduction of AI in networks with the help of multi-agent systems (MAS) constitutes a promising approach. The MASs present a new way of considering the resolution of a problem. The system's behavior is guided by the search for compromises between several thinking "intelligences", with different viewpoints, about a complex problem to resolve [LAB 93, WEI 99]. The "*agent-oriented*" approach makes it possible to build flexible systems with complex and sophisticated behaviors through the combination of modular components. The intelligent components (the agents) and their interaction capabilities constitute a multi-agent system, whose capabilities are superior to the simple sum of the capacities of individual agents.

Generally, an agent can be seen as a software element responsible for the execution of part of a process. It contains a certain level of intelligence ranging from simple predefined rules to self-sufficient AI machines. An agent can act in the name of a user or a process thus enabling the automation of a task.

In general, the term "intelligent agent" extends from adaptive user interfaces, also known as interface agents, to intelligent processes cooperating together to execute a common task (cooperating agents) [BRI 01]. These agents are able to bring their expertise to solve incidents and control the network infrastructures. Their properties, which are mainly autonomy, adaptation and distribution, respond well to problems of managing widely spread subnetworks. Their use offers a better adaptation of services to the needs of the user [BUC 00].

The main motivation that has led to the use of agents in networks is the need to automate control and management processes and to quickly adapt the network's behavior to clients' needs [AGO 00, MOU 03].

1.6.3.1. *Intelligent agent principles*

Several types of agents exist: software agents, intelligent agents, mobile agents or personal agents (also called *personal digital assistants*, PDA). The main software development companies that normally use object technologies are showing an increasing interest in agents. Three causes or sectors can explain this interest:

– AI: for a long time AI has provided complex and hard to develop software despite the influence of the object-oriented approach [LAB 93]. AI researchers have long understood the advantage of working with software entities that are simpler, more flexible and reusable. In [MUL 96], Muller says: "*the group of agents is more than the sum of the capabilities of its members*";

– mobility: after the integration of the object world in the network environment, the necessity to provide autonomy and mobility to objects has appeared. New object-oriented languages have been designed with this objective and the term "agent" appears with the following definition: "*an agent is a computer program that acts autonomously on behalf of a person or organization*" [OMG 97];

– networks: they use software "agents" for control. We must remember that these agents do not have close relationships with the current definitions of intelligent agents. The SNMP and CMIP agents are software modules which retrieve registry management information for one or several network elements and communicate this information on demand from a monitor through the SNMP or CMIP protocol. On the other hand, the expansion of networks has given a new direction to the traditional method of using computers: the resources are distributed among different heterogenous systems and the information is spread over different networks. The agent approach is a solution deemed reliable by companies such as Oracle for the definition of a new "client/agent/server" architecture.

1.6.4. *Example of agent-based telecommunications network monitoring architecture*

Using the model proposed in [MOU 03], we will briefly describe a multi-agent architecture for dynamic QoS management based on the TMN model previously described. This model has three levels:

– the SML (*service management layer*) level is represented by a group of agents called SMA-service. The SMA-service is made up of three agents:

- the interaction agent,
- the validation agent,
- and the service monitor agent [MOU 03];

– the NML (*network management layer*) level is represented by one or more SMA-PDP. Each multi-agent system is made up of three agents:

- the PDP agent,
- the network control agent,
- and the scheduler agent;

– the EML (*element management layer*) level is represented by one SMA-PEP in each network element. An SMA-PEP is made up of the two agents:

- the PEP agent,
- and the control agent.

All these agents throughout the different management levels control and manage quality of service policy rules for the user (end-to-end view). A translation module is used to translate service contract clauses into policy rules in the QPIM format for SMA-service, and in the QPIM format to PIB in the case of SMA-PDP. Figure 1.3 illustrates the general architecture of the system.

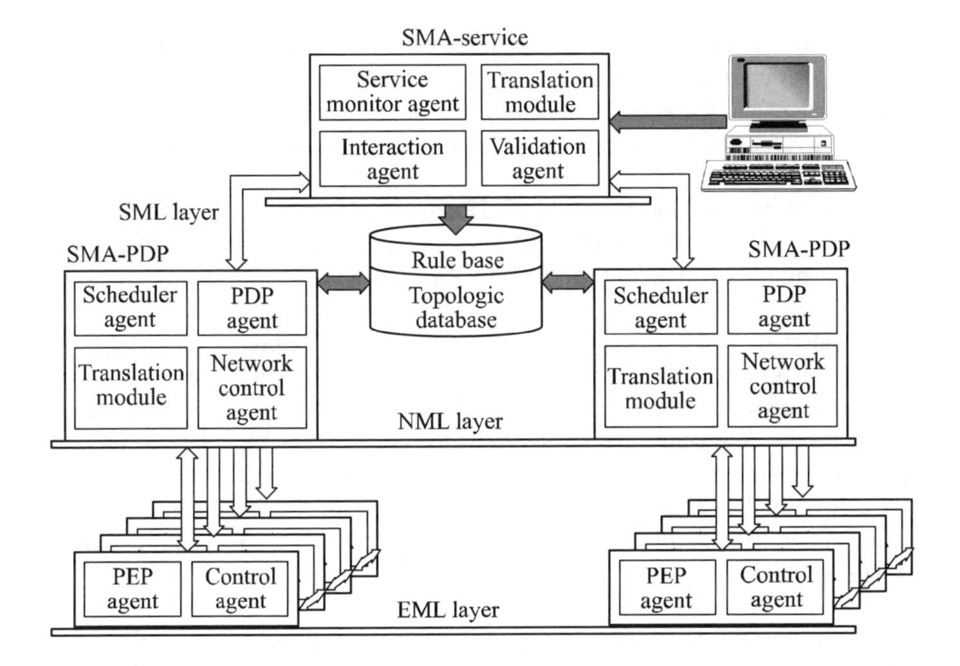

Figure 1.3. *General architecture of the system*

1.6.5. *Telecommunications network management with mobile agents*

1.6.5.1. *Overview*

Mobile agents (MA) can move from one node to another during their activity. Consequently, they can use the information collected during their migrations in order to adjust their behavior. In addition, they can move towards resources which are necessary for their activity. In the next sections, we will present how these properties can be used in telecommunications network management.

1.6.5.2. *Mobile agents*

According to [MOR 98, MAG 95], Magedanz presented solutions for the deployment of MAs which were intended for the management of telecommunications networks. Mobile agents encapsulate management scripts and send them where they are required. An agent can be sent into the telecommunications network and come back once it has collected management data. Sending an MA for this task replaces centralized monitoring operations. We should mention that the size of an agent must remain small in order to make efficient use of the bandwidth.

Similarly, a mobile agent can be used for local management operations. The operator can encapsulate these operations in a mobile agent and send them to the appropriate place. Since the agent makes the local decisions, there is no transmission of messages for this type of operation anymore, thus preserving bandwidth.

These ideas were implemented by FTP software [SOF 97]. The IP auditor is an application which sends mobile agents in the target network in order to collect management information, such as the configurations and state of network elements.

Another original and concrete application of MA is presented in [APP 94]. Mobile agents are used to command or control traffic congestion in a circuit switching network. A first class of MA called "parent agents", navigates randomly and collects information among the network nodes. They can then identify the congested nodes. When a congested node is identified, a *load-balancing* mobile agent is created in order to update the routing tables of the neighboring nodes (by using an optimal algorithm) and consequently decrease the traffic controlled by the congested node.

In a study of code mobility [BAL 97] presents the advantages of mobile agents. For example, if a network operator can only be connected through an uncertain or expensive link, he can create an *offline* MA, which only connects to the network to send the agent and reconnects later to retrieve the collected results. This principle is applied in Astrolog [SAH 97] to support "mobile managers".

[MAG 95] and [MAG 96] suggest other MA application possibilities. A network or service provider can send MAs to end-users in order to adapt their equipment to the new services. These agents can also execute other tasks such as user count and collecting requirements. [HJA 97] also suggests the use of MAs for quick personalization and immediate creation of services.

1.6.5.3. *Example of telecommunications network monitoring in the case of routing by ant colonies*

AI techniques also show all their power in the case of problems where the statement, data or parameters permanently change over very short periods of time.

This is the case for routing in telecommunications networks: when communication is established between two computers, the initial message is split into data packets which circulate along a network made up of transmission lines – whose throughput capabilities can be diverse and changing through time – and routers constituting the nodes of that network. The function of the routers is to direct data packets toward another router and so on and so forth until the data packets arrive at their destination. The router must take into account the traffic size over the communication channels to which it is linked in order to avoid a bottleneck. In this way, the data packets for one message may often follow different channels.

In order to avoid these bottlenecks the data packets travel with routing agents, virtual ants if you will, that analyze in real-time the state of congestion of the different network channels and indicate this state to each of the routers. The ants calculate the time it takes to travel from one node of the network to another and mark with a virtual pheromone the channel that they have taken. The shorter the delay, the more significant the intensity of the mark. Thus, when a data packet arrives at a given router, it will stand a better chance of using a channel in which the virtual pheromone is denser. In this way, the network adapts in real-time and in a decentralized way to the traffic activity [BON 00a, BON 00b, BON 99].

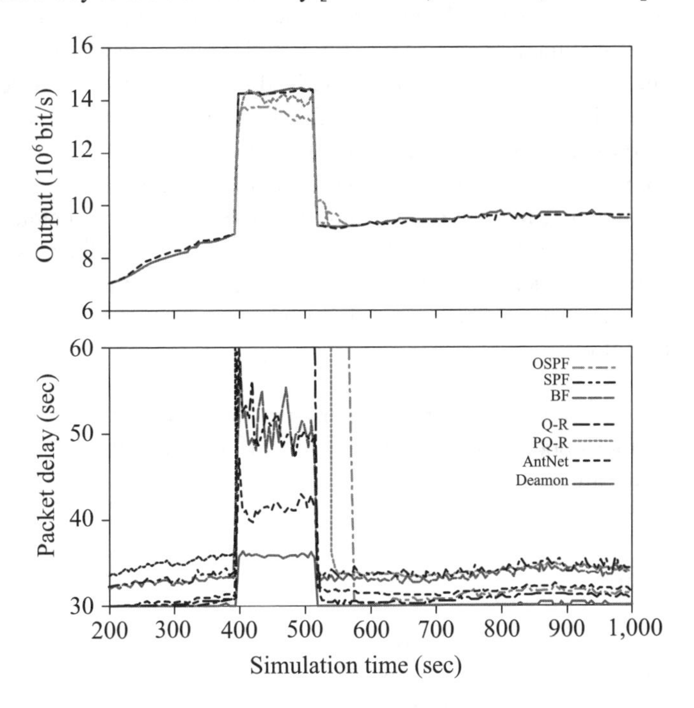

Figure 1.4. *Routing by ant colony [BON 00a, BON 00b, BON 99]*

Figure 1.4 represents a comparison between *AntNe*, a routing algorithm by ant colony [BON 99, BON 00a, BON 00b], and other routing algorithms. *Daemon* is an approximation of an ideal algorithm and offers a base for the best possible performance. The network is simulated in high traffic with stochastic traffic characteristics. At time t = 400, we simulate a sudden growth which lasts 120 seconds and takes the network to saturation. The top graph compares the network flow for the different algorithms, whereas the bottom graph compares the average delays of messages in a 5 s window. *AntNet* offers the same packet flow as the best algorithms while maintaining an average delay which is significantly lower than that of the other algorithms [BON 00a, BON 00b, BON 99].

1.7. Conclusion

In this chapter we have presented some recent studies concerning the application of AI in the field of telecommunications network management.

An intelligent system is only interesting when more simple solutions are not appropriate. Telecommunications network monitoring represents a prolific field for the application of certain AI techniques described earlier because of its complexity and the considerable growth of services offered by telecommunications networks. Using AI has provided a certain degree of automation and flexibility to monitoring approaches. With the help of AI-based tools, the human operator can act on a high level by using policies, motivations and goals. Furthermore, AI techniques provide detailed information on the state of the network's elements and their use, and can authorize corrective operations and proactive actions. For these reasons, many companies have invested in these AI techniques and a few are participating in the standardization efforts.

The future of AI in telecommunications networks is based on the needs of telecommunications operators for coping with management, control and monitoring problems and for improving their services.

1.8. Bibliography

[AGO 00] AGOULMINE N., FONSECA M., MARSHALL A., "Multi domain policy based management using mobile agent", *Computer Science*, 2000.

[APP 94] APPLEBY S., STEWARD S., "Software agents for control", in Cochrane P. and Mealthey P. (ed.), *Modelling Future Telecommunication Systems*, Chapman & Hall, 1994.

[AZA 93] AZARMI N., CORLEY S., AZMOODEH M., BIGHAM J., PANG D., "Model based diagnosis in maintenance of telecommunication networks", *Proceedings of the International Workshop on Principles of Diagnosis(DX'93)*, p. 46-59, Aberystwyth, 1993.

[BAL 97] BALDI M., GAI S., PICOO G.P., "Exploiting Code Mobility in Decentralized and Flexible Network Management", *Proceedings of the First International Workshop on Mobile Agents*, Berlin, April 1997.

[BON 99] BONABEAU E., DORIGO M., THERAULAZ G., *Swarm Intelligence: from natural to artificial systems*, Oxford University Press, 1999.

[BON 00a] BONABEAU E., DORIGO M., THERAULAZ G., "Inspiration for optimization from social insect behaviour", *Nature*, vol. 406, p. 39-42, July 2000.

[BON 00b] BONABEAU, E., THERAULAZ, G., "Swarm Smarts", *Scientific American*, 282 (3), p. 72-79, 2000.

[BRI 01] BRIOT J.P., DEMAZEAU Y., *Principe et architecture des systèmes multi-agents*, Hermes, Paris, 2001.

[BRU 96] BRUSONI V., CONSOLE L., TERENZIANI P., THESEIDER DUPRÉ D., "A spectrum definition for temporal model based diagnosis", *Proceedings of the International Workshop on Principles of Diagnosis (DX'96)*, p. 44-52, Val Morin, France, 1996.

[BUC 00] BUCKLE P., DINIS M., CUTHBERT L., "Distributed intelligent control and management for 3G networks", *Nortel Networks PLC London*, Portugal Telecom Inovacao, Electronic Engineering, 2000.

[CAS 94] CASSIOPEIA, "OSA systems: structure, functionality and behavioral description by means of a distributed processing model", *A project of the RACE program Ref: 2049*, January 1994.

[CAI 94] CASSIOPEIA, "Integrated service engineering: OSA validation report", *A project of the RACE program Re: 2049*, January 1994.

[CAS 95] CASSIOPEIA, "Integrated service engineering: OSA validation report", *A project of the RACE program Ref: 2049*, January 1995.

[CON 91b] CONSOLE L., TORASSO P., "A spectrum of logical definitions of model based diagnosis", *Computational Intelligence*, p. 133-141, 1991.

[CON 91b] CONSOLE L., TORASSO P., "A spectrum of logical definitions of model based diagnosis", *Computational Intelligence*, p. 133-141, 1991.

[DBE 95] DE BELER J., KOWALCZYK A., LECKIE C., ROWLES C., "AI for managing telecommunication networks", *Worldwide Intelligent Systems*, p. 87-108, 1995.

[DKL 87] DE KLEER J., WILLIAMS B.C., "Diagnosing multiple faults", *Journal of Artificial Intelligence*, vol. 32, no. 1, p. 97-130, 1987.

[DOL] ACTS-DOLMEN, "Service machine for an open long-term mobile and fixed network environment", *ACTS Ref: AC036 Dolmen.*

[EUR 91] EURISTIC SYSTÈMES, *Chronos, outil de développement de systèmes experts temps réels*, 1991.

[HAM 92] HAMSHER W., CONSOLE L., DE KLEER J., *Model based diagnosis*, Morgan Kauffmann Publishing, p. 1-520, 1992.

[HJA 97] HJÁLMTÝSSON G., JAIN A., "An Agent-based Approach to Service Management towards Service Independent Network Architecture", in *Integrated Network Management: integrated management in a virtual world*, p. 715-729, San Diego, California, IFIP, Chapman & Hall, May 1997.

[LAB 93] LABIDI S., LEJOUAD W., "De l'intelligence artificielle distribuée aux systèmes multi-agents", *INRIA* no. 2004, August, 1993.

[MAG 03] http://www.telecom.gouv.fr/rnrt/projets/res_01_48.htm.

[MAG 95] MAGEDANZ T., "On the Impacts of Intelligent Agents Concepts on Future Telecommunication Environments", *Third International Conference on Intelligence in Broadband Services and Networks*, Greece, October 1995.

[MAG 96] MAGEDANZ T., ROTHERMEL K., KRAUSE S., "Intelligent Agents: An Emerging Technology for Next Generation Telecommunications?", *INFOCOM'96*, p. 464-472, USA, March 1996.

[MAY 99] MAYER E., Apprentissage inductif de scénarios pour la supervision de réseaux de télécommunications, PhD Thesis, University of Rennes 1, 1999.

[MOR 98] MORSY M.C., PIERRE C., JACQUES L., "Intelligent Agents in Network Management a State-of-the-Art", *Networking and Information Systems*, vol. 1, no. 1, p. 1-29, 1998.

[MOU 03] MOUNGLA H., KRIEF F., "For an Intelligent Policy-based Telecommunication Networks Management", *International Conference of Telecommunications Management Networks*, 2003.

[OMG 00] OMG, "OMG white paper, Open service Marketplace", version 1.0, http://www.omg.org, May 2000.

[PAR 95] PARKYN N., "Architecture for Intelligent (Smart) Agents", http://www.citr.uq.oz.au, June 1995.

[PFA 93] PFAU WAGENBAUER M., AUSTRIA S., NEJDL W., "Integrating model-based and heuristic features in a realtime expert system", *IEEE Expert*, vol. 8, p. 12-18, 1993.

[REI 87] REITER R., "A theory of diagnosis from first principles", *Journal of Artificial Intelligence*, vol. 32, no. 1, p. 57-96, 1987.

[RIE 93a] RIESE M., "Diagnosis of extended finite automata as a dynamic constraint satisfaction problem", *Proceedings of the International Workshop on Principles of Diagnosis (DX'93)*, p. 60-73, Aberystwyth, 1993.

[ROZ 99] ROZÉ L., Supervision de réseaux de télécommunication: une approche à base de modèles, PhD Thesis, University of Rennes 1-IRISA, 1997.

[SAH 97] SAHAI A., BILLIART S., MORIN C., "Astrolog: A Distributed and Dynamic Environment for Network and System Management", *Proceedings of the 1st European Information Infrastructure User Conference*, Germany, http://www.irisa.fr/solidor/doc/pub97.html, February 1997.

[SAM 01] SAMMOUD K., Ingénierie du déploiement des services de télécommunication sous contraintes de QoS, PhD Thesis, ENST, 2001.

[SEB 98] TRIGILA S. *et al.*, "Mobility in long-term service architectures and distributed platforms", *IEEE Personal Communication*, August 1998.

[SOF 97] SOFTWARE F., "FTP Software Agent Technology", *Technical Report*, FTP Software, http://www.ftp.com/product/whitepapers/4agent.htm, 1997.

[STA 96] STALLINGS W., *SNMP, SNMPv2 and RMON, Practical Network Management*, Addison-Wesley, USA, 1996.

[TER 92] TERPLAN K., *Communication Networks Management*, Prentice Hall, 1992.

[TIN] TINA CONSORTIUM, "A cooperative solution for a competitive world", http://www.omg.org.

[UIT 96] UIT-T, "Recommandation M.3100: principe des réseaux de gestion des télécommunications", May 1996.

[UIT 00] UIT-T, "Recommandation M.3010: TMN and network maintenance: international transmission systems, telephone circuit, telegraphy, facsimile and leased circuits", 2000.

[UIT 01] UIT-T, *Recommandation Q1201: Réseau intelligent – principe de l'architecture.*

[UIT 02] UIT-T, *Recommandation Q1202: Réseau intelligent – architecture du plan de service.*

[UNG 93] UNGAUER C., Problématique d'utilisation de techniques de supervision à base de connaissances profondes: L'exemple de la supervision du réseau TRANSPAC, Technical report, CNET, October 1993.

[WEI 99] WEISS G., *Multi-agent Systems: A Modern Approach to Distributed Intelligence*, MIT Press, 1999.

[ZNA 97] ZNATY S., GERVAIS M.P., *Les réseaux intelligents: ingénierie de services de télécommunications*, Hermes, Paris, 1997.

Chapter 2

Adaptive and Programmable Management of IP Quality of Service

2.1. Introduction

The new Internet generation is characterized by the coexistence of several types of media in one infrastructure. This convergence has motivated the development of new capabilities for the Internet protocol (IP), which was not initially designed to receive this concentration of media types.

However, the coexistence of all these types of media can lead to serious network management problems. Different traffic profiles, each with their own requirements and priorities which sometimes share limited resources are considered among them. The introduction in the network infrastructure of services with constraints which have been based until recently on the philosophy of best effort has prompted the development of quality of service (QoS) support. This development consists of characteristics required by an application to ensure good data reception quality for the user.

In short, a network infrastructure has QoS support when it is able to process and respect the different constraints linked to its applications. However, providing QoS over IP is not an easy task. New architectures such as IntServ and DiffServ [WAN 01] have been developed for meeting these requirements. The underlying mechanisms making this QoS support possible (mainly consisting of queue

---

Chapter written by Miguel CASTRO, Dominique GAÏTI, Abdallah M'HAMED and Djamal ZEGHLACHE.

management and scheduling) have recently been the subject of very popular research. For this reason, several propositions have come forth in various works relative to these two tasks. It is important to mention that the development of each new algorithm is designed to resolve the existing weakness in a specific situation. Consequently, there is a significant variety of traffic management mechanisms and making the right choice can be difficult.

If we consider that each traffic management mechanism has its own strengths and weaknesses, we could intuitively conclude that if the network node knows how to classify traffic conditions, then it could apply the best management configuration known for this situation, which could optimize the network's performance. This constitutes the main motivation for our study. We propose an adaptive approach for QoS management to improve network performance. In this perspective, we present a new network management architecture which is able to carry out the adaptive management of QoS. This architecture is characterized by a management plane based on active and programmable network technology. This new management plane can execute internal control and monitoring cycles in order to adopt the adaptive behavior.

Furthermore, this architecture remains open to receive online behavior updates. The programmability of the management plane is based on a new programming language called CLAM (*compact language for adaptive management*). This chapter also presents three case studies where CLAM architecture is used to improve QoS performance with adaptive management.

This chapter is divided into eight sections. Section 2.2 presents the active and programmable network technology which will serve as framework for the proposition in this chapter. In section 2.3, we describe the adaptive and programmable approach for QoS management in IP networks. Section 2.4 presents the CLAM architecture developed to achieve this new adaptive QoS management. The new language developed for this architecture is presented in section 2.5. In section 2.6 we present other studies and approaches on the subject, and we compare these with the CLAM approach. Section 2.7 illustrates three case studies where adaptive QoS management was used to improve network performance. Finally, section 2.8 presents the conclusions and perspectives on these research studies.

2.2. Open and programmable network technology

The almost static environment of traditional network infrastructures makes the deployment of new services difficult. The hardware infrastructure must be adapted in order for the new service or protocol to be deployed. Consequently, the idea to add more processing power within network nodes was presented to provide the necessary flexibility to facilitate the dynamic development and deployment of services, and at the same time to enable the introduction of new, more intelligent management services.

Researchers in this field are divided into two main communities: the "active networks" community supported by DARPA and the "OpenSig" community introduced by Columbia University in the USA. These two communities centralize projects which belong to several research institutes throughout the world. The points of view concerning the level of desirable flexibility may be different, but the final goal is the same for both communities: the creation of a network infrastructure that is flexible for an easier deployment of new services and protocols.

The idea of having more flexible network infrastructures came about at the same time as the need for QoS in the Internet and motivated the creation of new traffic management mechanisms. Even though the normalization efforts from the Opensig group (IEEE P1520 project [BIW 98]) did not produce the expected results, the idea of a complete separation of planes which make up the Internet's functional model (data or forwarding plane and control plane) inspired the Forces (*forwarding and control elements separation*) group [YAN 04]. All these efforts support a higher level of flexibility in the architectures and functionalities of the network, from the way the packets are processed in the network to more sophisticated functions such as resource reservation and routing. This separation leads to the possibility of designing and implementing network functions in an independent way, thus enabling, for example, performance improvements obtained by hardware execution in data transport, while providing software flexibility for the implementation of new control and management services.

In order to support this type of flexibility, new hardware technologies have been proposed. ASIC (*application-specific integrated circuit*) type circuits are not adapted to the renewed network service and protocol requirements. We are then experiencing the appearance of an alternative for these ASICs, called network processors (NP). These processors are specifically designed for processing packets which can increase the performance of software routers.

2.3. Active and programmable QoS management over IP

In [CAS 04b], we have shown that the configuration of QoS management may not be appropriate for all the possible traffic scenarios. We can come up with specific traffic scenarios in which the performance obtained by applying a given traffic management configuration may be inadequate when compared with those obtained with other scenarios. Besides, there are situations where complex QoS management mechanisms show no increase in performance when they are compared to more simple mechanisms. For example, Christiansen *et al.* [CHR 01] have shown that the RED mechanism does not show any performance improvement when compared with the FIFO mechanism in the case of web traffic. Le *et al.* [LE 03] arrived at the same conclusion. Therefore, the early rejection of packets, as imposed by the RED mechanism, is not justified in this case. In addition, the RED mechanism can also increase the problems of equity on the aggregation of reactive and non-reactive flows [LIN 97]. The relation between implementation complexity and performance must also be considered in the choice of a good QoS management configuration.

The network must be able to recognize current traffic conditions intuitively and must adapt to them in order to increase performance. However, this adaptation task is not as simple as a traditional management function. The unpredictable and dynamic behavior of traffic in the network requires a real-time operation to make swift decisions and adapt quickly. For this task, the traditional SNMP-based management with a central administrator analyzing several managed objects is completely inappropriate. Round-trip time (RTT) required by SNMP messages makes this operation impossible [SCH 00].

It is therefore sensible to equip network entities with the capability of some self-management, thus bringing decision-making closer to the managed objects. This capability will reduce management reaction time and give the network more autonomy.

In traditional management, the network elements have limited processing power and no decision control on their own management. All decisions are generally taken by the administrator from information collected by network polling or by interpreting control messages. The traditional network management procedure illustrated in Figure 2.1 is handled within a centralized administrator.

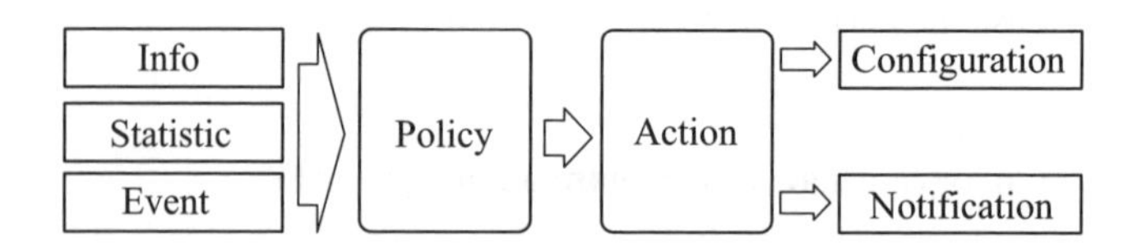

Figure 2.1. *Traditional management procedure*

The idea behind the adaptive management architecture is an execution of specific simple management procedure occurrences shown above within each element of the network. In addition, the goal is to maintain an architecture that is open for "on the fly" deployment of new management behaviors.

The architecture presented in the next section was designed to ensure an adaptive management based on a modular point of view. Each component in Figure 2.1 is translated into a programmable management module. The interaction between these modules results in the behavior that the network element will execute, which means that the management procedure illustrated in this figure is integrated in the network element instead of being executed in the management center which often relies on human intervention.

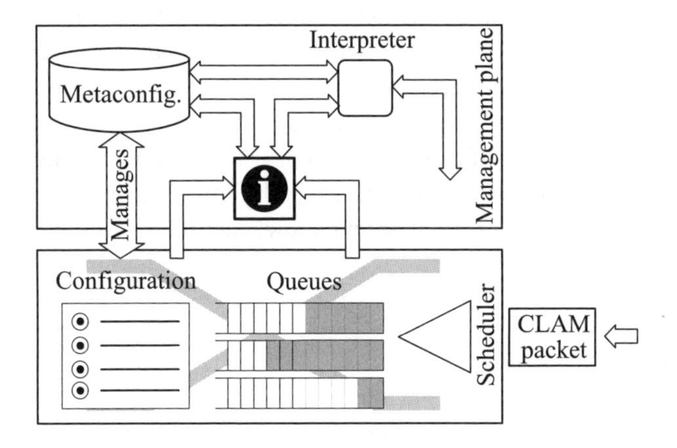

Figure 2.2. *Simplified functional model of the execution environment*

Figure 2.2 illustrates a simplified functional model dedicated to adaptive management. Generally, the adaptive management plane is an independent entity associated with the network element. In this figure, the adaptive management plane is located over the forwarding plane. In the management plane, the information is collected and processed. Based on this knowledge, the management plane will act on the forwarding plane by modifying its configuration.

The behavior of an equipment set up with adaptive management is defined by its metaconfiguration (a term that will be discussed later in this chapter). A metaconfiguration is the result of an interaction between multiple modules. For the modeling of a metaconfiguration, these modules must be implemented and the knowledge of the entity must be defined and modeled. This knowledge is necessary to control the relation between programmed modules. For example, the knowledge base must define the action to take when a given notification arrives, or the configuration to modify when a given threshold is reached.

The knowledge base is explained by a set of rules and can be developed in different ways. It can be a representation of mathematical models and in this case, the rules can represent an algorithm executing a given task. For example, the rules can be modeled to conduct an adaptive approach of the RED mechanism as proposed by Vukadinović and Trajković [VUK 04], in which formulae are used to update RED parameters for the purpose of an optimization.

The knowledge can also be obtained from the observation of a real network or from simulations. In the first case, network traffic reports can be used to extract knowledge. For example, these reports can be used to identify events which can deteriorate the network's performance, and policies can be created to control and execute tasks to avoid future problems. However, the observation of a real network can be complicated. It is difficult to figure out whether certain actions which must be executed when a given event occurs are really effective or not. In this context, network simulation is the most interesting knowledge source for modeling adaptive management knowledge.

2.3.1. *Programmable modules*

The architecture described in this section is based on the programmability of several modules. A set of programmed modules describes a *metabehavior*. The relation between all these modules is illustrated in Figure 2.3. "Event" and "Action" are the main components of a metabehavior. The other components are programmable modules interacting between each other according to actions and events. Adaptive management knowledge can be explained by ECA (event-condition-action) rules.

Programmable modules within adaptive management are explained in more detail in Figure 2.3.

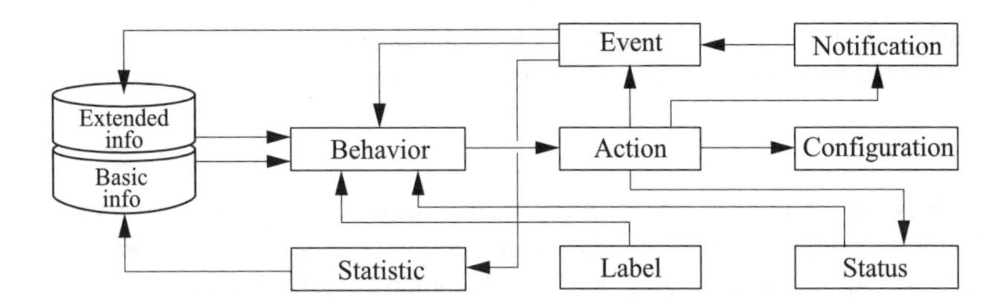

Figure 2.3. *Programmable adaptive management modules*

2.3.1.1. *Information*

The traditional management operation based on periodic requests (*polling*) produces information in sometimes excessive and superfluous quantities, which must be transported by the network to a management center. This leads to a waste of bandwidth. Even with voluntary transfer technologies (*pushing*), where information is sent to the administrator without an explicit request, the decision on information transfer requirements is based on simple conditions and no measure can be taken to react to potential problems from the object control side. In addition, traditional centralized processing of management information complicates scaling for larger domains.

It is possible to process management information locally within the network's elements now that the processing capacity is less expensive. In addition, this can reduce useless traffic in the network. Furthermore, the decision can take into account more specific and abundant information analysis, such as the contents of higher packet layers, which would be complicated with centralized management.

2.3.1.2. *Statistic*

A statistic is a special type of continuing information where its value is updated by a designated procedure. As opposed to the information type previously described, the statistic module can launch an inherent procedure in order to update its value. Also, as with information, a statistic becomes a criterion in decision-making. Other programmable modules can access its value by its identifier.

2.3.1.3. *Status*

Network entities can have configurable statuses. This improvement can lead to the quick evaluation of the node's state and the personalization of operational statuses. For example, we can define a status within the nodes of a network in order to determine the congestion level in the surrounding area. The difference between "status" and "information" modules is that status changes characterize events. A status change can launch actions, which is not the case with information.

2.3.1.4. *Label*

Labels can be allocated to network entities in order to choose the appropriate behavior for each network entity class. For example, we can assign a label to a node in order to define if it is "edge", "core" or "interdomain edge". This simple attribute can be used, for example, to decide if an incoming programmed module meets its function. It would then be possible to decide whether a module should or should not be installed in the node.

Compared to the programmable "information" module, the main difference is that a label generally represents global static information, whereas information can be volatile.

2.3.1.5. *Configuration*

The definition of occurrences in the programmable "configuration" module enables policies to launch configuration changes more easily. The configuration module contains all the procedures which are required to be executed in the forwarding plane in order to activate/deactivate management mechanisms.

2.3.1.6. *Notification*

Personalized notifications such as programmable modules can be defined in the architecture. Each node can produce and send notifications to terminal systems, surrounding nodes, or to the network management center. The semantics and syntax of each note can be configured. This module replaces *push* technologies. The difference between *push* technology and the programmable notification module is that *push* mechanisms are only related to the management center. On the other hand, notifications can also be sent to other network entities. For example, a node can send congestion notifications to its neighbors in order to launch cooperative mechanisms to minimize or even reduce congestion effects. Receiving a notification launches an event and then the behaviors relative to this event will be executed to process this notification.

2.3.1.7. *Behaviors*

Behaviors are defined in order to determine the node's operation. Specifically, these modules command adaptive management behavior by policies. The behaviors are based on conditions which are imposed by other programmable module occurrences such as statistic, information or status (Figure 2.4).

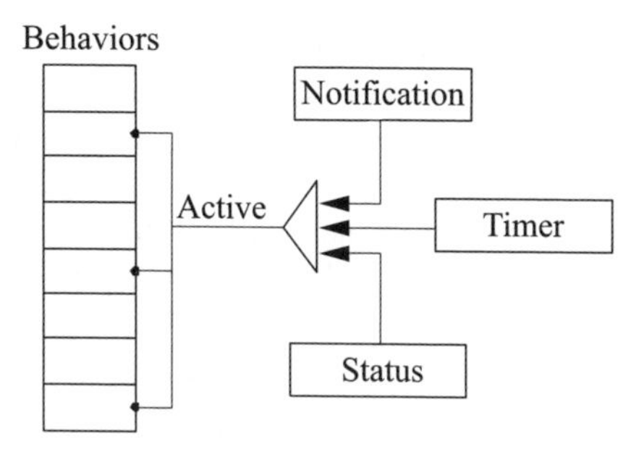

Figure 2.4. *Behavior activation agents*

The knowledge base used as a source for adaptive management is represented mainly by a set of behavior type modules. A specific number of behaviors are installed in a network entity and each behavior is launched in response to a given event. A behavior can be launched by a notification event, a status event or a timer event. Whereas the first two types of events depend on other programmable modules, the third follows a time frequency which is defined within the behavior. For example, we can create a behavior to launch each second to adapt RED thresholds on a queue based on a certain formula.

2.4. Architecture for adaptive and programmable management

In order to accomplish the adaptation task for current traffic conditions, network management must have enough processing power to be able to carry out basic control and monitoring functions. In addition, it is prudent to keep management architecture open in order to support the deployment of new management behaviors at all times. New technologies for open and programmable networks offer the necessary support for the deployment of these means.

In other words, active and programmable network technology presents all the necessary characteristics to generate adaptive and programmable management of QoS.

We introduce the term "behavioral network management" to describe network management in which the nodes are generally responsible for their operation and maintenance in order to execute adaptive management actions or, more precisely, actions associated with QoS and real-time control. This relative autonomy is vital in order for the network entities to evaluate their environment and modify their behavior according to their findings. These findings which determine the best behavior for a given scenario are designated as *metabehaviors* in this chapter. In adaptive network management, the behaviors are translated into configurations and therefore the term *metaconfiguration* is considered as a synonym of metabehavior.

We propose the inclusion of a management plane in the Internet's layered architecture in order to reach our goal of achieving adaptive capabilities. The type of structure in three-dimensional layers is based on the B-ISDN (broadband integrated services digital network) reference model [ITU 92] which includes a distinct plane where all management functions are implemented.

A separation of functions is also considered in the Forces workgroup of IETF [GRO 02]. According to Forces, forwarding and control functions can be separated and the communication between these two planes is based on a series of standardized protocols. In addition to these two planes, the management plane can also be separated and can execute its own tasks independently and collaboratively.

Our approach is different from traditional management functions because we propose a management plane based on activity and programmability in order to arrive at open and adaptive management. This new management plane will retain traditional management mechanisms and it will also be able to arrange new behaviors by providing a management API and a specialized execution environment. The architecture proposed for the active management plane is shown in Figure 2.5. With these new capabilities, the management plane is able to carry out *in band* management tasks (directly integrated in forwarding functions) as well as *out of band* tasks (requiring management protocols and administrator/agent communication).

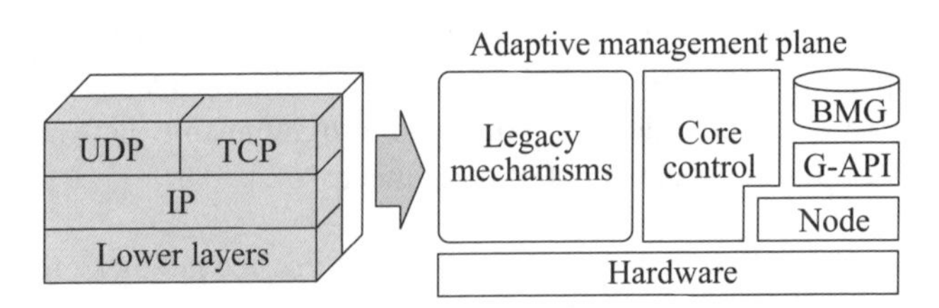

Figure 2.5. *Active and programmable management architecture*

The adaptive management plane interacts with the two other operational planes which make up the functional model: control plane and forwarding plane (*data*). The control plane is responsible for control functions such as routing and resource reservation. The forwarding planes take care of the way packets are processed and forwarded. Since control plane functions could potentially be very sophisticated

(BGP, RSVP, etc., for example), the forwarding plane processing overload must be minimized and the separation of functions between control and forwarding planes is necessary. As an example, the simplest forwarding function must read the incoming packet header, interrogate the routing table and forward the packet towards a given outgoing interface. This task is done by the forwarding plane, whereas the control plane is responsible for creating and maintaining the routing table.

The management plane relies on a modular view of the forwarding plane as proposed in the *DiffServ router* management model [BER 02]. In this model, the DiffServ router is activated by the interoperability between modules such as dropper, classifier, scheduler, etc. This modularity is important to enable an easy access to the forwarding functions.

The adaptive management plane illustrated in Figure 2.5 is made up of the system's hardware which is shared between two entities: the set of legacy mechanisms and the management kernel. In the upper part of the management kernel there is the core control of the architecture, the management application programming interface (MAPI) and the management modules base (MMB). These three modules make up the adaptive management execution environment (AMEE).

2.4.1. *Legacy mechanisms*

Legacy mechanisms are incorporated mechanisms which are inherent to the network architecture chosen. The RED, WRED, ECN, WFQ, etc. mechanisms are all examples. In reality, these mechanisms represent configurations or an implementation of blocks which are built from the underlying forwarding plane as described in the *DiffServ router* management, as well as management algorithms such as SNMP and COPS which are largely deployed on the Internet and are vital to retain the interoperability between traditional and active nodes. These mechanisms basically integrate the normalized implementation which must be available to the network entity in order to support the interoperability with entities inherited from the network such as "*best effort*" routers.

2.4.2. *MMB*

The MMB is a permanent memory which stores available personalized management modules used by adaptive management. The MMB updates are totally controlled by the core control element and, logically, by central transport management entities. Since metabehavior is obtained from the association of several programmable modules, we can conclude that MMB stores the metabehaviors installed in the node.

Access to MMB is reserved for authorized entities, which means that the deployment of personalized modules within the architecture is not accessible to each end-user. Therefore, the general definition of active networks in which each user is able to inject and execute codes in the network is not used in this adaptive management architecture. On the other hand, this architecture can be viewed as being closest in similarity to the philosophy of programmable networks (*OpenSig*).

2.4.3. *MAPI*

The MAPI represents all the available functions and procedures to form metabehaviors. It constitutes the interface between the management plane and the representation of the physical router, which is based on the *DiffServ router* management model [BER 02].

Among available functionalities within this API, we find functions for modifying parameters in a given module, activating and deactivating a module or creating and injecting packets in the network.

2.4.4. *Management kernel*

The management kernel is an operating system responsible for creating and processing the adaptive management execution environment (MEE) dedicated to core control and to all management models built over the MAPI and stored in MMB. The *DiffServ router* management model relies on this module.

In terms of implementation, the management kernel can be integrated in a dedicated calculator and communicate with the forwarding plane through a normalized protocol. Nevertheless, it is vital for the management kernel to have high speed access to the forwarding plane in order to ensure real-time control.

2.4.5. *Core control*

The role of the core control is to manage the activation of installed modules, to monitor the network state and to enable the network administrator to decide, based on installed policies (thanks to programmable modules), whether to replace current modules with available inactive modules, according to the observed state of the network. The core control is also responsible for providing management information to the management support system.

2.4.6. *Hardware*

This module represents the hardware platform through which the management execution environment is deployed. As proposed by the Forces workgroup for control and forwarding planes, the adaptive management plane can also be deployed within an independent unit, such as a separate box, card or even a processing plane in a multiprocessor. In this case, the processor scheduler must maintain the independence between the different planes in order to avoid interprogram overloads.

2.5. CLAM: a new language for adaptive and programmable management

The adaptive metabehaviors of network elements are expressed in a new domain-specific language (DSL) called CLAM (*compact language for adaptive management*). This language is specifically designed for writing all types of programmable modules dedicated to the architecture discussed earlier. CLAM is a simplified language with limited syntax and does not enable loops or memory pointers. Nevertheless, CLAM syntax can express all necessary operations for the management plane, even for real-time metabehaviors (such as QoS configuration allocation) and for traditional management functions (accounting management, for example).

The choice of not using loops or memory pointers within the CLAM syntax is justified by the security advantages gained. The limited syntax has already been used in active networks as a security mechanism and has shown good results [HIC 98]. However, the lack of loops within a programming language limits its power of expression and its efficiency. Consequently, this absence in CLAM must be compensated in other ways.

CLAM codes can use functions and procedures defined by MAPI, including access to the information base of the device (in a MIB SNMP, for example) and calls for procedures and management functions. Since the definition of modules in CLAM can contain references to other modules (for example, status, information, action, etc.), a dependence control must be executed in order to accept the activation of new modules. More information on this programming language and on the implementation of this architecture on the OPNET Modeler™ simulation tool [OPN 04] can be found in [CAS 04a].

2.6. Related studies

In this section, we present a few studies relative to the dynamic and autonomous management of networks. Some of them use an active network approach and others are based on different dynamic approaches.

2.6.1. *Behavioral networks*

Merghem *et al.* [MER 02] have proposed the use of multi-agent technology to control QoS in a dynamic network environment. This architecture, which is represented by behavioral networks, is based on a behavioral network model where nodes are represented by agents and where different behaviors, such as "prudent", "confident", "not confident", can be attributed to each node (agents). These behaviors define the manner in which passing traffic must be controlled by the node. Based on its own knowledge, an agent can modify its behavior in order to face its environment (incoming traffic). On the other hand, the cognitive agent is responsible for controlling reactive agents by activating or deactivating them. The way in which the cognitive agent controls its reactive agents is dictated by its metabehavior.

Generally, CLAM and behavioral networks have the same goals. Both architectures attempt to use adaptive and dynamic network management in which the network is partially responsible for its own management. Consequently, the term "metabehavior" is also used in CLAM architecture and generally has the same significance in both architectures. The main difference between them resides in the approach used for adaptive behavior integration in the network. In fact, the multi-agent domain and the active network philosophy have much in common. However, we believe that active networks can provide more flexibility in network management.

2.6.2. *Smart packets*

Schwartz *et al.* [SCH 00] have proposed a network management architecture called "smart packets" and based on active networks. This approach is based on a compact code transported through the network and executed in distant nodes to carry out network management tasks such as SNMP GET and SET commands. Instead of simple information collection requests resulting from traditional management, smart packets also carry the way in which the collected information must be processed and the actions that will need to take place in the future. The results of the packet's execution, if they exist, are sent back to the management center. As with CLAM, smart packets are encapsulated with the ANEP protocol.

The approach of the smart packet architecture is similar to CLAM. It does not attempt to maintain the permanent code in the nodes. This characteristic removes the possibility of programming the permanent code dedicated to the creation of different behaviors in each node. In addition, the possibility of installing behaviors within a node and of using it indefinitely decreases the number of safety and security controls. In CLAM, the code must be verified only once during installation and can be easily activated and deactivated.

Smart packets are specifically designed to take into account managed objects and agents in SNMP. All the decisions made by smart packets are therefore based on MIB variables. CLAM goes beyond that by supporting the creation of mid- to long-term information, other than what is contained in MIB. The responsibility of the management center in smart packet architecture is still very complex. Since the nodes are only able to evaluate the codes when they arrive, no management task is done because the management center will not have injected active packets.

2.6.3. *SENCOMM*

Jackson *et al.* [JAC 02] have proposed an architecture for management based on active network and called SENCOMM (*smart environment for network control, monitoring and management*). As with "smart packets", SENCOMM is based on the execution of compact programs in nodes through ANEP packets. Even if it is possible to install a permanent code in the SENCOMM management execution environment (SMEE), this operation is more adapted to the installation of libraries. In SENCOMM, management operations are carried out by so-called "smart probes", which are executable codes transported in ANEP packets. These smart probes use available libraries in the execution environment, such as API SNMP, to collect information or to execute actions. The execution environment used for SENCOMM is the active signaling protocol (ASP EE) [BRA 02], which is based on the Java programming language. Smart probes in the SENCOMM prototype must also be implemented in Java.

The management of the SENCOMM execution environment coexists with other execution environments within the active node. The approach chosen for this architecture is the philosophy of completely active networks in which users can inject and execute codes in active nodes. The CLAM approach differs from SENCOMM by the fact that the programmable and active network domain is limited by the management plane. In addition, the use of Java as programming language, as with ABLE architecture, can bring unnecessary complexities to the network management task.

2.6.4. *General evaluation*

The approaches presented above use active networks in order to provide flexible and effective network management. However, these approaches may not be able to support the behavior programmability or flexibility of nodes at the same level as CLAM. The main difference between CLAM and approaches such as smart packets and SENCOMMM is that CLAM remains close to the forwarding plane with the *DiffServ router* management model. The information collection is directly conducted from the forwarding plane. Nevertheless, access to information by SNMP still remains possible. The other approaches are mainly based on the SNMP view of network management, even if this management is carried out close to or inside managed objects. CLAM goes beyond by trying to provide network management tasks that SNMP GET and SET commands are unable to do. These tasks require more proximity with management and forwarding planes, while keeping a certain independence between them, especially in terms of performance. We can therefore consider that CLAM targets the access to the architecture's lower layers that SNMP cannot reach.

In comparison with the other architectures, the CLAM architecture was designed on the basis of an approach that is different from the programmable and active networks domain. CLAM is closer in similarity to the OpenSig community (programmable networks) compared to the DARPA active network philosophy. In other words, whereas the other approaches are generally based on packet processing behaviors such as receive-verify-execute-destroy or receive-verify-execute-forward, CLAM adopts the receive-verify-install behavior. It is for this reason that the execution of programmable modules in CLAM is self-controlled and can be launched by external players with the help of simple authorized notifications.

CLAM is more realistically intended for QoS management of packet configuration. In fact, the need for adaptation of QoS configuration at this level is the main motivation of CLAM. The case studies presented in the next section show that CLAM is in fact very effective for this task. In addition to operating in lower layers, the architecture can also be used for traditional management tasks with SNMP.

2.7. Case studies

In this section, we present a few case studies in which the use of QoS adaptive management architecture helps us reach three different goals. In the first case study, we use adaptive management to specifically optimize web service performance. In the second case study, we change the adaptive behavior goal by defining a more complex objective function where several performance metrics are considered. The third case study uses the CLAM architecture to control the equity problem between TCP and UDP flows in a congested IP network based on best effort. The architecture was implemented in a commercial network simulation tool (OPNET Modeler™ [OPN 04]). Details on the development process of the simulated architecture can be found in [CAS 04a].

2.7.1. *Case study 1: web service optimization*

The behavior and composition of traffic have a strong influence in the performance of traffic management mechanisms. In [CAS 04b], we have shown that the relative quantity of UDP flow in aggregated traffic directly influences the choice of a good traffic management configuration.

The goal for this case study is to analyze the behavior of three queue management configurations when they are used by aggregated traffic, which is made up of TCP and UDP flows. We model and examine a metaconfiguration where the node adapts to known conditions by analyzing current traffic conditions in order to associate the queue management mechanism that is best suited for the situation.

2.7.1.1. *Scenario and metaconfiguration specification*

The topology used as the basis for our experimentations is illustrated in Figure 2.6. This topology is made up of two router nodes, a set of IP servers and 170 clients of which 150 statically use a set of applications based on the TCP protocol and 20 use services based on an *on-off* UDP. The number of active UDP clients will vary over time. Clients and servers are at opposite ends of a link connecting both routers with a bandwidth of 20 Mbps.

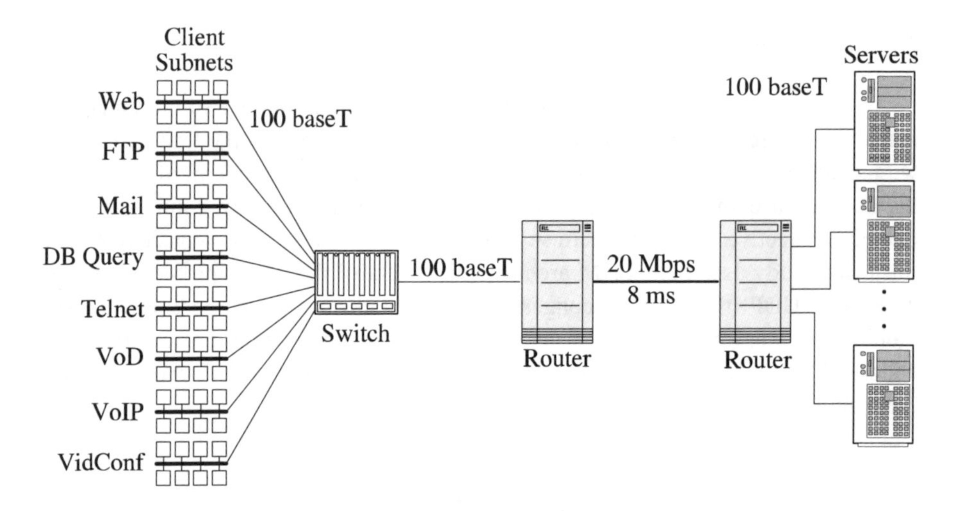

Figure 2.6. *Simulated topology*

In this case study, the *DiffServ router* model associated with our adaptive IP router is illustrated in Figure 2.7.

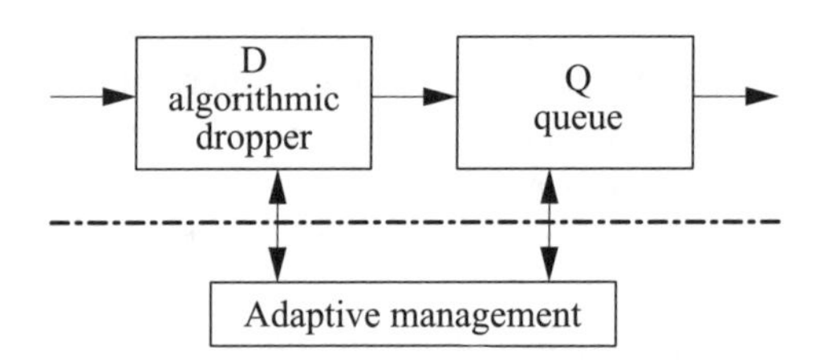

Figure 2.7. *DiffServ router model for case studies 1 and 2*

The profiles chosen for the queue management are: FIFO with *drop-tail* (*FIFO*), FIFO with RED (*RED*) and FIFO with RED and the addition of the ECN (*explicit congestion notification*) mechanism [RAM 01]. Then, we have defined several different scenarios where UDP traffic is variable. We examined situations where UDP traffic varies from 5 to 85% in relation to the total traffic. For each UDP traffic ratio, we have measured the performance of all three queue management configurations.

We have then defined a method to estimate the instant ratio of UDP traffic over aggregated traffic of an interface by packet sampling. Another (more precise) way of obtaining this information could be to question the SNMP agent in the router. However, this evaluation by packet sampling enables us to examine the "statistical" programmable module operation provided by the CLAM architecture.

Every 4 ms we collect a sample (packet clone). The identifier of the transport protocol used is stored in a continuing vector of 1,000 positions (last 1,000 samples). Then, every 10 s, the basic metabehavior is activated to analyze the sampling data of the last 4 s and make the appropriate decision based on the current situation. The evaluation of the UDP traffic ratio based on these parameters has returned results that were sufficiently precise with respect to actual values.

From consecutive simulations on the behavior of the three defined profiles we can specify a metaconfiguration that will attribute the correct profile to use for a known UDP traffic ratio in order to reach our goal.

Figure 2.8 shows the profiles to use with respect to the UDP traffic ratio in aggregated traffic.

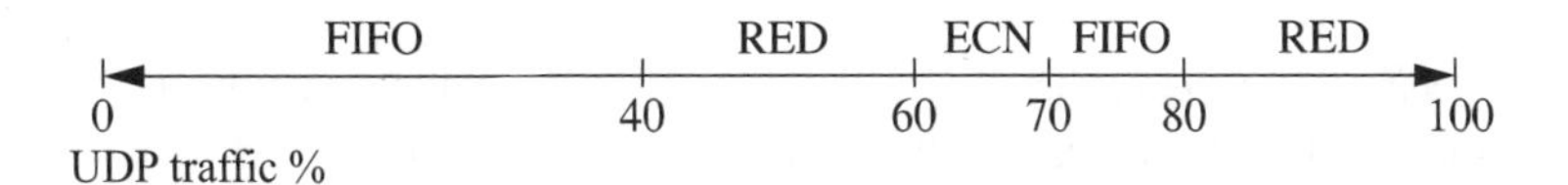

Figure 2.8. *Profiles activated to minimize the loading time of a web page*

The basic operation of our metaconfiguration is to switch between all three queue management configurations:

– FIFO;

– RED; and

– ECN.

This operation is achieved by simply changing the parameters of the "D" module on the *DiffServ router* model in Figure 2.7.

The CLAM source code corresponding to the three configuration modules and the statistic module through which the configuration allocation decision will be made are shown below in Figure 2.9.

```
define config FIFO
as
iface.D.max_p : = 0 ;
iface.D.ecn_status : = 0 ;
end
define config RED
with float new_max_p ;
as
iface.D.max_p : = new_max_p
 ;
iface.D.ecn_status : = 0 ;
end
define config ECN
with float new_max_p ;
as
iface.D.max_p : = new_max_p
 ;
iface.D.ecn_status : = 1 ;

   end
```

```
define info last_1000pkts
type int[1000]
end
define info ndx type int end
define statistic udp_load type int
 clock 0,004 as
int proto_id ;
IPPacket pkt ;
collectPacket (pkt,iface.q1) ;
proto_id : = getTranspProtId(pkt) ;
info.last_1000pkts[info.ndx] : =
 proto_id ;
info.ndx ++ ;
if info.ndx > = 1000 then
info.ndx : = 0 ;
done
update(hashCount(info.last1000pkts
 17)) ;

   end
```

Figure 2.9. *CLAM source code for queue and statistic configurations*

The vector of integers called `last_1000_pkts` is used to retain the identifiers of the transport protocol of the last 1,000 packet samples. The `udp_load` statistic counts all the positions with a value of 17 corresponding to the UDP protocol indentifier.

If we consider the `udp_load` statistic as the main decision making criteria, we can model the findings extracted from preliminary simulations (Figure 2.8) in a programmable "Behavior" type module, called `casestudy1`, which is launched every 10 s (Figure 2.10).

```
define behavior casestudy1 prior 0 trigger clock 10 as
if udp_load < 400 then
 activate(FIFO) ; else
 if stat.udp_load > = 400 && stat.udp_load < 600 then
 activate(RED, 0,1) ;
 else if stat.udp_load > = 600 && stat.udp_load < 700
then
 activate(ECN, 0,1) ;
 else if stat.udp_load > = 700 && stat.udp_load < 800
then
 activate(FIFO) ; else activate(RED, 0,1) ;
 done
 done
 done
done
sleep 10 ;
end
```

Figure 2.10. *CLAM source code of behavior module*

In order to test the metabehavior's possibilities of using the modeled knowledge and to verify its effectiveness in reducing web page loading time, we have created a random scenario where the number of active UDP clients changes over time. The active router does not know these variations beforehand. Therefore, the statistical method by sampling is supposed to estimate these changes. All the other simulation results will now be based on this new random scenario.

2.7.1.2. *Results and discussion*

For each defined scenario, the simulation was repeated a certain number of times (with various random generation sources) in order to obtain a representative average. For example, during web page upload, we obtained an average value of $\mu = 0.7802$ s and a confidence interval of $\pm$ 0.059 s for $\alpha = 0.05$. The duration of each simulation is 670 s and the analysis does not take into account a preliminary transient period of 130 s, which is determined by the method described in [JAI 91]. We have used the same simulation procedure for the next two case studies.

We have then compared the performance of all three defined profiles (FIFO, RED and ECN) with the performance reached from an adaptive profile (shown in the results as "*active*") when they are applied to the previously mentioned random scenario and where the UDP load changes over time. The metaconfiguration defined for the *active* profile has forced queue management adaptation. The adaptation behavior of queue management for this experiment is illustrated in Figure 2.11. This illustration shows the current queue management configuration for every 15 s of simulation. It is interesting to note that the queue management configuration has changed to adapt to current traffic conditions.

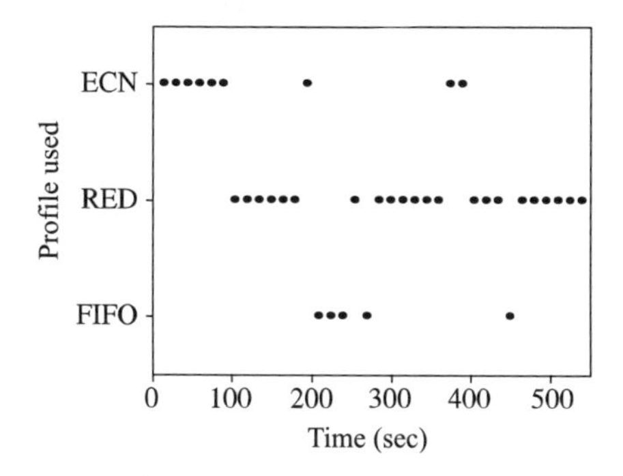

Figure 2.11. *Typical behavior of QoS configuration for case study 1*

Figure 2.12 represents samples of the results obtained. It shows the progress of web services (Figure 2.12a) and database (Figure 2.12b) performances for each profile. With these results where the goal is to reduce the average uploading time for a web page, we can observe that among all the configurations examined, the *active* profile has generated the best average uploading time. This means that the objective pursued for this experiment was met. With respect to the database service performance, the metabehavior for this experiment (Figure 2.8) coincides with the best adaptation profile to also minimize the response time for a database request. This can be explained by the fact that the traffic generated by both applications has similar characteristics. Consequently, the adaptation necessary to optimize an application coincides with the other application.

In Figure 2.13, we present a global performance assessment of all four profiles with respect to a predetermined goal for this experiment. This figure shows that the *active* profile has in fact the shortest average for web page uploading time, which satisfies our objective.

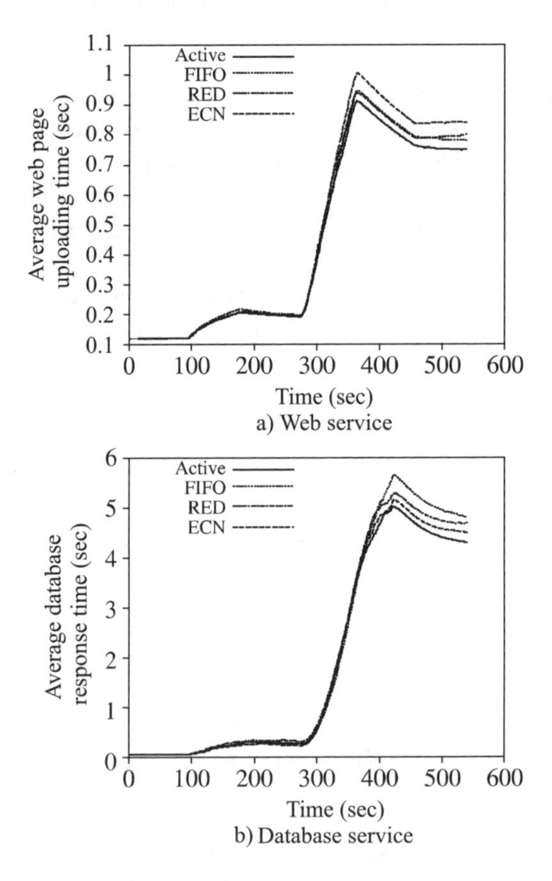

Figure 2.12. *Web services and database performance for case study 1*

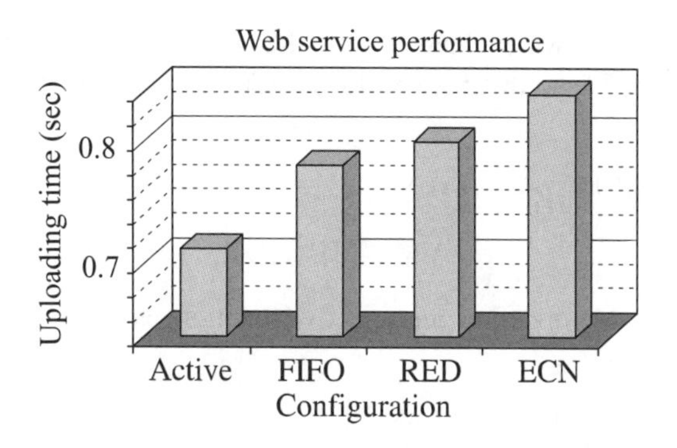

Figure 2.13. *Web service uploading time for all configurations*

2.7.2. *Case study 2: maximization of a given objective function*

For this case study, we want to use adaptive management to minimize an arbitrary objective function, which is defined by performance metrics from services and network. We will use the same structure as the last case study.

We also retain the simulation topology, queue management configurations, *DiffServ router* model (Figure 2.7), client behavior and estimation of UDP load by sampling. The only change is our goal.

2.7.2.1. *Scenario and metaconfiguration specification*

In order to establish the desired objective function we have defined 10 criteria including network performance measures (packet loss ratio and average queue occupation) and performance of applications (HTTP/web, database query, FTP, Telnet, videoconference and voice over IP). We attribute a weight (w_c) to each criterion representing its importance in the evaluation of the objective. The criteria and their respective weights are given in Table 2.1.

Criterion	**HTTP/web**	**BD**	**FTP**	**Telnet**	**Packet loss**
Weight	6.0	4.0	1.0	1.0	2.0
Criteria	File occup.	VC delay	VC jitter	VoIP delay	VoIP jitter
Weight	0.7	1.0	0.2	1.0	0.2

Table 2.1. *Criteria and their weights*

For each criterion c we evaluate a normalization factor among the performance measures P_c obtained from the three profiles: FIFO, RED and ECN.

$$F_c = \mathrm{Max}\left(\bar{P}_c^{FIFO}; \bar{P}_c^{RED}; \bar{P}_c^{ECN}\right)$$

For each profile M we define our objective function, called "relative gain", as:

$$G^M = \frac{1}{\sum w_c} \times \sum_{c=1}^{C}\left[w_c \times \left(1 - \frac{\bar{P}_c^M}{F_c}\right)\right] \qquad [2.1]$$

where w_c is the weight attributed to criterion c. The goal of this experiment is to maximize the function given by equation [2.1].

Again, with the help of consecutive simulations, we have obtained different performance metrics corresponding to the three queue management profiles and for the scenarios with increasing UDP traffic ratio. By comparing the results obtained from the profiles examined and by choosing the profile with the best performance for each UDP load we have defined the desired behavior for adaptive management. This behavior is presented in Figure 2.14. Consequently, based on the knowledge generated by the simulations, if the node adapts its queue management configuration to the behavior shown in the illustration with respect to the estimated UDP load, the resulting value of the objective function will be maximized.

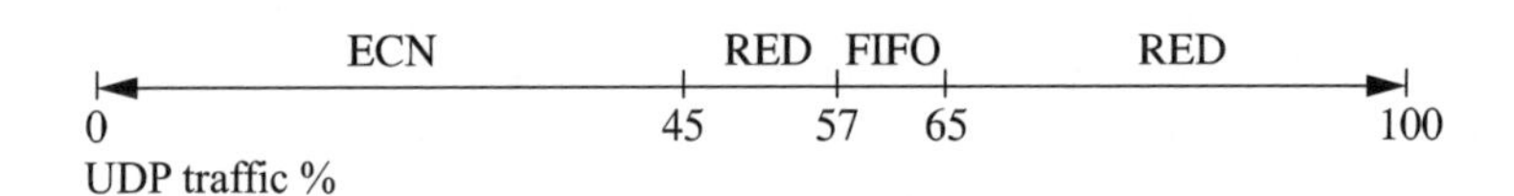

Figure 2.14. *Profiles activated to maximize the objective function*

CLAM modeling for this experiment is very similar to that used in the first case study. The only exception is the programmable module of the behavior type which must be adapted to the new objective. The new behavior is shown in Figure 2.15.

```
define behavior casestudy2 prior 0 trigger clock 10
as
if udp_load < 450 then
 activate(ECN, 0,1) ;
else
 if stat.udp_load > = 450 && stat.udp_load < 570
then
 activate(RED, 0,1) ;
 else
 if stat.udp_load > = 570 && stat.udp_load < 650
then
 activate(FIFO) ;
 else
 activate(RED, 0,1) ;
 done
 done
done
sleep 10 ;
end
```

Figure 2.15. *CLAM source code for behavior module in case study 2*

CLAM programmable modules have been uploaded in the router node and we have used a random scenario, as specified in the last case study, to examine the effectiveness of the adaptation to an unknown situation where the UDP load must be estimated.

2.7.2.2. *Results and discussion*

In Figure 2.16 we show the behavior resulting from the adaptive configuration of queue management in the node. This figure shows that in the last experiment, the configuration behavior of queue management has changed over time to adapt to traffic fluctuations. Clearly, the change profile is different from the first case study because the goal has changed.

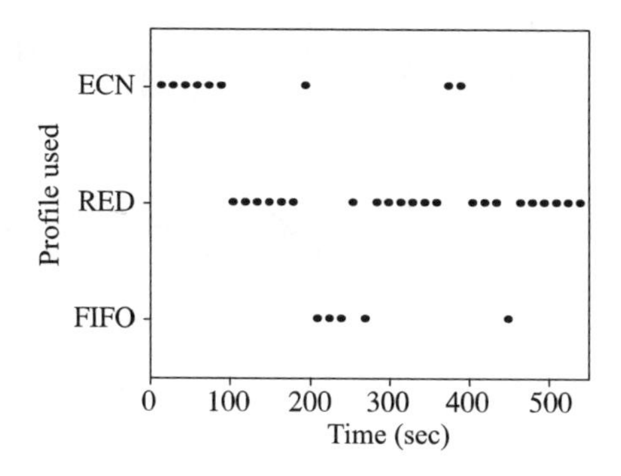

Figure 2.16. *Typical behavior of QoS configuration for case study 2*

Figure 2.17 illustrates the comparison between average uploading times of a web page obtained on studied profiles. We can see that the *active* profile performance is not the best among all the others. However, since the objective function defined also takes into account the performance indicators from other criteria, this result is not enough to confirm that the objective function has not been maximized.

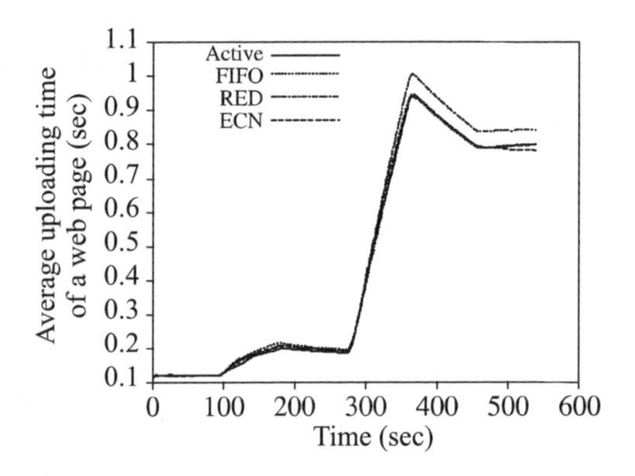

Figure 2.17. *Web service performance for case study 2*

We now compare the global performance of the four profiles with respect to the objective indicated for this experiment, i.e. the values of the objective function (equation [2.1]). In Figure 2.18 we can see that the *active* profile provided the best value for the objective function, which satisfies the specified objective for this case study.

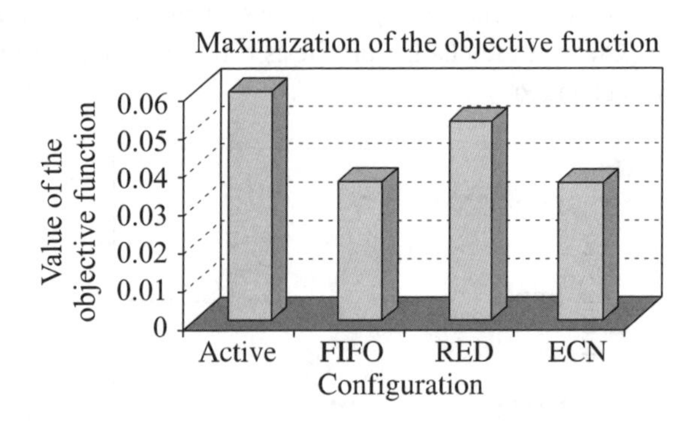

Figure 2.18. *Objective function values for all the configurations*

2.7.3. *Case study 3: adaptive control of equity*

In our third case study, the CLAM architecture is used to conduct an adaptive control of packet scheduling parameters in order to decrease equity problems among UDP and TCP applications in a congested network. In a congested node the non-reactive behavior of UDP flows can be harmful to the coexistence with TCP flows which are considered as having a "good behavior". In fact, a UDP flow cannot adapt its throughput during congestion, or in other words it is not "TCP-*friendly*". So in case of congestion UDP flows tend to deprive TCP flows, thus generating an equity problem. This case study uses adaptive management to decrease the magnitude of this problem.

2.7.3.1. *Scenario and metaconfiguration specification*

We use the same topology (Figure 2.6) and applications as in the last two case studies. First we will analyze the equity problem phenomenon with a simple experiment: 300 TCP clients are competing with a variable number of UDP clients for the 20 Mbps connection. Each UDP client generates a constant bit rate (CBR) of 910 kbps. In the router we have an interface with only one queue which has a capacity of 100 packets and is configured in RED (*random early detection*) [FLO 93], with the following parameters: $Max_{th} = 100$; $Min_{th} = 50$; $p_{max} = 10$. We then calculate the equity index for this scenario as defined in [JAI 91]:

$$F(x_1, x_2, \cdots, x_n) = \frac{\left(\sum_{i=1}^{n} x_i\right)^2}{n \times \sum_{i=1}^{n} x_i^2}; \quad x_i = \frac{T_i}{O_i} \qquad [2.2]$$

where T_i and O_i respectively represent the measured and fair throughputs for flow i.

We have observed the equity index in two preliminary cases. In the first case, the basic *DiffServ router* model used is the same as that of the first two case studies, i.e. a *best-effort* router with a simple queue. In the second case, we use the basic *DiffServ router* model of Figure 2.19.

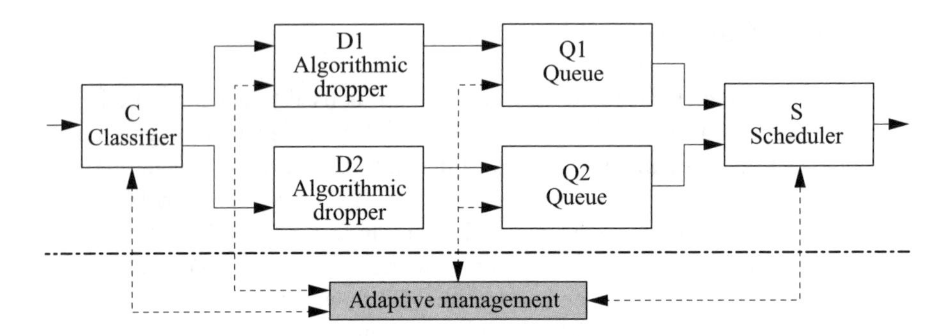

Figure 2.19. *DiffServ router model for case study 3*

The “C” classifier is configured to choose packets based on their “Protocol” field in the IP header. Packets based on UDP (value 17) must be sent to the “D1” dropper and all the others are forwarded to the “D2” dropper. Consequently, “D1” will only receive UDP packets, whereas “D2” receives all the other packets. We use the CQ (*custom queuing*) mechanism [VEG 01] as a scheduling algorithm for the module “S” of the *DiffServ router* model. At first, we want to evaluate the impact on equity of simple traffic separation in two different queues, with similar priorities, without – for now – adapting the parameters. We have defined two queues (“Q1” and “Q2”) with a 50 packet capacity each. The two algorithmic droppers (“D1” and “D2”) are modeled with RED mechanism and are configured as: $Max_{th} = 50$; $Min_{th} = 30$; $p_{max} = 10$.

In order to calculate the equity index, we take into consideration the fair throughput for an UDP flow as being the data generation CBR throughput, whereas for TCP flows we consider the fair throughput as being the maximum consumed throughput by a TCP flow when there is no competing UDP flow. Figure 2.20 shows the results of these two simulations.

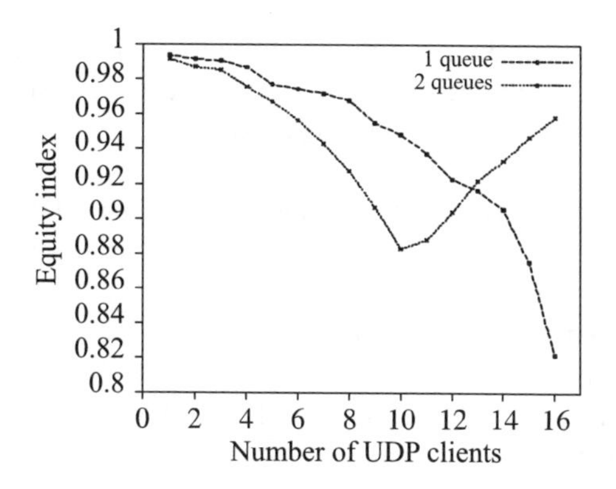

Figure 2.20. *Equity index values for one or two queues*

In Figure 2.20 we can observe that, in the one queue configuration, the equity index calculated with equation [2.2] decreases when the number of UDP flows increases. Quantitatively, during UDP load increase, we have observed that the TCP flows must generally decrease their throughput by 62%, whereas the UDP flows reduce their throughput on average by only 11% because of packet rejects. With two queues, as expected, we observe that the equity index decreases as long as the number of UDP clients has not reached 10, then it starts to increase. This is due to a parity of priority levels which we defined between both queues. 10 average UDP clients represent 10 Mbps of throughput which corresponds to 50% of the outgoing connection capacity. Since, from 10 UDP sources, there is no available bandwidth for UDP traffic anymore, UDP packets start to be rejected at each full queue and then, the equity index starts to go up again.

Although we have been able to observe a certain improvement of the equity index due to traffic separation into two queues, the static bandwidth sharing – allocating 50% of resources for each queue – is not yet considered as fair. Consequently, it can be desirable to control equity in an adaptive way in order for the loss ratio of both types of traffic to be proportionally equivalent during congestion, which would increase equity between these types of traffic. This adaptive control can be achieved by CLAM architecture metabehavior.

For this case study, equity control metabehavior is based on a simple heuristic: we must observe (by sampling) the arrival of packets (UDP and TCP) and then determine the appropriate parameters for the scheduling algorithm *custom queuing*, i.e. the *byte counts* for both types of traffic. As with the previous case studies, we collect packet samples to determine the UDP load. We collect a sample every $\Delta\tau$ ms. After assembling information from N packets, new values for CQ *byte count* parameters are calculated. Then, a decision is taken at each $N \times \Delta\tau$ ms, based on the UDP traffic rate among the last N packets. After each decision, all counters are zeroed and the counting procedure is launched again.

In order to determine the best connection between the UDP traffic rate and the CQ configuration we have put in place several simulations in which multiple UDP load values have been tested. For each situation, different CQ configurations have been examined. From the analysis of the results we have been able to find the appropriate CQ configuration profile to increase equity based on the UDP traffic relation. By a simple linear regression we have found the γ_{tcp} and γ_{udp} factors representing the relation of the total *byte counts* (BC_{all}) given to TCP and UDP queues respectively. Consequently, $\gamma_{tcp} + \gamma_{udp} = 1$. These factors then depend on TCP (R_{tcp}) and UDP (R_{udp}) traffic rates on packet samples. In order to protect ourselves against possible queue privation phenomena, we have decided to determine a minimum of 0.025 for each factor. In other words, each queue will have at least 2.5% of *byte counts*. *Byte count* values for both queues are given in equation [2.3]:

$$BC_{tcp} = \begin{cases} \gamma_{tcp} \times BC_{all} & \text{if } \gamma_{tcp} \geq 0.025 \\ 0.025 \times BC_{all} & \text{if } \gamma_{tcp} < 0.025 \end{cases}$$
$$BC_{udp} = \begin{cases} \gamma_{udp} \times BC_{all} & \text{if } \gamma_{udp} \geq 0.025 \\ 0.025 \times BC_{all} & \text{if } \gamma_{udp} < 0.025 \end{cases} \qquad [2.3]$$

Table 2.2 shows all the simulation parameters defined for our experiments. As with previous case studies, the transient simulation time was calculated based on the techniques from [JAI 91]. For the same scenario, several simulation series have been conducted in order to achieve a reasonable confidence level close to the level found in the first two case studies.

Parameter	Name	Value
Total simulation time (s)	T_{sim}	430
Transient time (s)	T_{trans}	130
Total *byte count*	BC_{all}	20,000
Number of collected packets	N	1,000
Time between collections (ms)	$\Delta\tau$	(1, 2, 4, 8, 16, 32, 64)
Number of UDP clients	N_{udp}	Variable (1 → 20)
Number of TCP clients	N_{tcp}	300

Table 2.2. *Simulation parameters*

From the results of preliminary simulations, we have determined the correct values for γ_{tcp} and γ_{udp} in order to calculate CQ parameters every Δt_d ms. Thus, the core control of the adaptive management kernel was configured with a metabehavior where we observe that the incoming traffic and CQ parameters are adapted in real-time. In order to examine the effectiveness of this metabehavior we have used the same random scenario as that from the other case studies where the number of active UDP clients varies over time. The core control does not have a direct access to information such as the instant number of active UDP clients. The UDP traffic relation in aggregated traffic must be estimated from collected packet samples. The contents of CLAM programmable modules, which are vital to the development of our metabehavior are provided below.

```
define statistic udp_count type int clock 0.008 as
int count;
IPPacket pkt;
collectPacket (pkt,iface.s);
 /* This time packet sample is collected from scheduler
*/
proto_id: = getTranspProtId(pkt);
count: = stat.udp_count;
if proto_id = 17 then
 count: = count + 1;
done
update(count);
end
```

Figure 2.21. *Definition of the statistic which counts UDP packets*

The programmable module from Figure 2.21, as those used in the first two experiments, retrieves a packet sample every 8 ms from the classifier (and not from the queues this time) and increments the counter if the packet corresponds to the UDP transport protocol. In fact, we have tried this "statistic" module with several sampling frequency values, as is shown in Table 2.2. Once more, there are other ways to evaluate this statistic. For example, we can integrate in the *DiffServ router* model two surveyor type blocks just before the two droppers ("D1" and "D2"). Another way would be to send a request to the router's MIB SNMP. However, the goal of this case study is to use and examine the different types of programmable modules. The other necessary programmable module is "behavior", as specified in Figure 2.22. This behavior calculates, with equation [2.2], the new *byte counts* for both queues, attributes them to the CQ scheduler and zeros the counting statistic of UDP packets.

```
define behavior casestudy3 prior 0 trigger clock 8
as
float new_udp_bc;
new_udp_bc: = (stat.udp_count/1000) * 20000;
if new_udp_bc < 500 then /* 2.5% x 20000 */
 new_udp_bc: = 500;
if new_udp_bc > 19500 then /* 97.5 % x 20000 */
 new_udp_bc: = 19500;
s.bytecount[0]: = new_udp_bc; /* UDP queue */
s.bytecount[1]: = 20000 - new_udp_bc; /* TCP queue
*/
stat.udp_count: = 0;
sleep 8;
end
```

Figure 2.22. *Behavior definition which updates Byte Count parameters*

2.7.3.2. *Results and discussion*

This section shows the results obtained from routers integrating the metabehavior described in the last section. In Figure 2.23 we can see the average UDP load estimated by the metabehavior and the corresponding γ_{udp} values for a scenario where $\Delta t = 4$ ms, i.e. a packet examined every 4 ms. These γ_{udp} values have been used to calculate the real-time adaptation for BC_{tcp} and BC_{udp} values. We can therefore confirm in this figure that γ_{udp} is directly proportional to the UDP traffic rate.

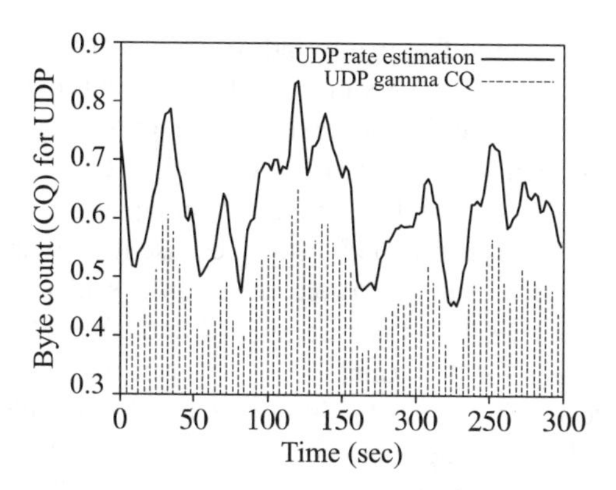

Figure 2.23. *The value of γ_{udp} with respect to the UDP rate estimation*

In order to determine the best ΔT_d value, we have tried different values for $\Delta\tau$ by setting the number of examined packets before decision with N = 1,000. We have tried the following ΔT_d = (1 s; 2 s; 4 s; 8 s; 16 s; 32 s; 64 s) values. Figure 2.24 shows the average value of equity index calculated for three of these scenarios (*active* and γ_{udp} = 1 s; 8 s; 32 s). We have also examined the two older scenarios where there is no adaptive management: 1 queue and 2 queues.

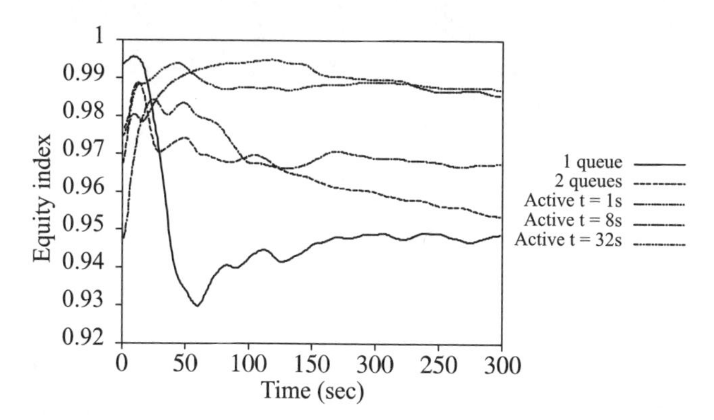

Figure 2.24. *Equity index evolution for different ΔT_d values*

As this figure shows, all the scenarios examined with active metabehaviors present equity improvements, compared with both set configurations (1 queue and 2 queues). Figure 2.25 compares the average equity index values for all scenarios. We notice that when we isolate all the active scenarios we get the lowest equity index with $\Delta T_d = 8$ s. Nevertheless, this equity index value is still better than with static scenarios. This can mean that the 8 s interval does not return a sufficiently precise value for the UDP traffic rate for the next 8 s. In fact, this estimation depends on global traffic behavior and specifically on its variability. On the other hand, the best values are found for $\Delta T_d = 1$ s and $\Delta T_d = 32$ s. Consequently, we conclude that the decision granularity of 1 s returns good results because it accompanies all the minor traffic behavior changes and the decision often consists of updates. On the other hand, $\Delta T_d = 32$ s is the choice that is best adapted to the behavior and variability of the modeled applications for this simulation. However, the $\Delta T_d = 1$ s configuration is not interesting because it requires the information to be collected from a packet at each millisecond, which in the worst case scenario would mean collecting the equivalent of 12 Mbps of data (presuming that MTU = 1,500 bytes). Consequently, this would lead to a prohibitive processing management cost. The most appropriate choice for our network behavior is $\Delta T_d = 32$ s.

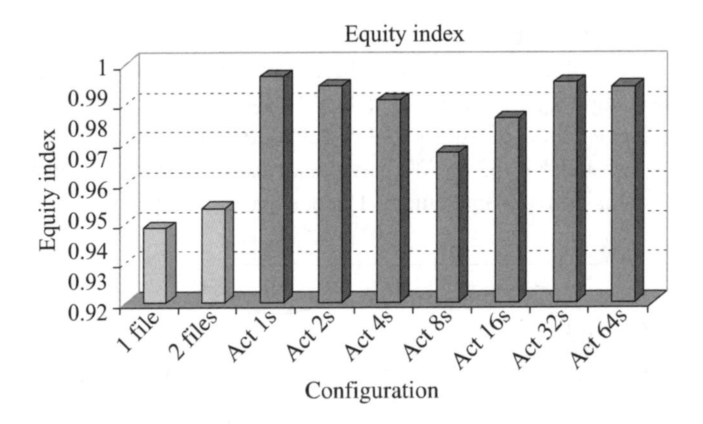

Figure 2.25. *Average equity index values for all the scenarios*

The observation of equity indices in Figure 2.25 shows that the gap between the best value obtained with the active metabehavior ($\Delta T_d = 32$ s) and the value obtained with the basic scenario (1 queue) is only 0.038. Although this may seem like a small difference, the impact on the final performance of the application is very significant. As an example, we have sampled the average web page uploading time for some scenarios. The results are illustrated in Figure 2.26. These results show a restoration of the web application offering a performance of approximately 20% when we compare scenarios $\Delta T_d = 32$ s and “1 queue”. In that case, the average uploading time for a web page decreased from 6.81 s to 5.47 s on a route with only one hop.

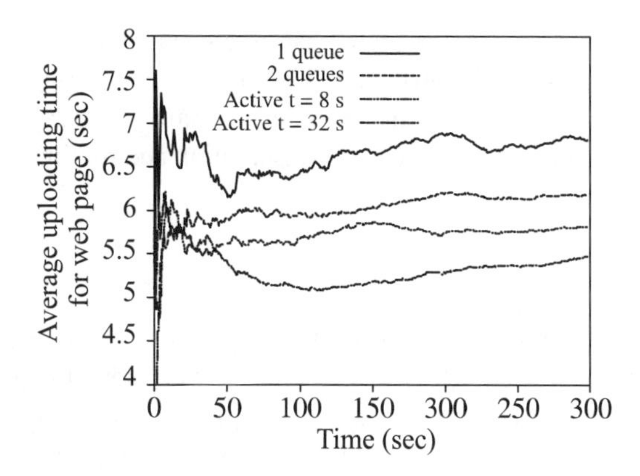

Figure 2.26. *Web service performance*

2.8. Conclusion and perspectives

This chapter presents an adaptive approach designed for network management and particularly QoS management. We propose a new architecture based on active and programmable network technology. This architecture's design was motivated by the possible performance improvements of an IP network when its QoS configuration can be adapted "on the fly", by using the strengths of each available algorithm in QoS support. Three case studies using this architecture are also presented in order to illustrate the capabilities and effectiveness of this adaptive approach.

The case studies have demonstrated encouraging results for the adoption of the adaptive behavior of QoS support. We have shown that the simple solutions to known problems can benefit from adaptive network management by increasing the performance.

The implementation of the architecture in a network simulation tool helps us to understand and test the characteristics of this architecture and its capacity to personalize its behavior "on the fly". Even if the implementation was done with a simulation tool, we think that the simulated IP router model used as the basis for our implementation is close to reality. In addition, the implementation of the CLAM language interpreter was based on simple Linux tools which enable us to easily move this implementation to an open network platform such as the Click modular router [KOH 01]. This constitutes the next step in this study.

We are also working on creating new case studies in which the potential for new programmable modules will be analyzed. We are also considering the specification of an interface with the control plane in order to enable the adaptive management plane to allocate functions such as resource reservation and routing.

2.9. Bibliography

[BER 02] BERNET Y., BLAKE S., GROSSMAN D., SMITH A., "An Informal Management Model for Diffserv Routers", *RFC* no. 3290, Internet Engineering Task Force, May 2002.

[BIW 98] BIWAS J., LAZAR A.A., HUARD J.F., LIM K., MAHJOUB S., PAUL L.F., SUSUKI M., TORSTENSSON S., WANG W., WEINSTEIN S., "The IEEE P1520 Standards Initiative for Programmable Network Interfaces", *IEEE Comm. Mag.*, vol. 36, no. 10, p. 64-70, October 1998.

[BRA 02] BRADEN R., LINDELL B., BERSON S., FABER T., "ASP EE: An Active Network Execution Environment", *Proc. of DARPA Active Networks Conference and Exposition (DANCE'02)*, San Francisco, USA, p. 238-54, May 2002.

[CAS 04a] DE CASTRO M.F., Gestion programmable et adaptive de la qualité de service sur IP, PhD Thesis, Ecole doctorale SSTO/Institut national des télécommunications, Evry, France, September 2004.

[CAS 04b] DE CASTRO M.F., MERGHEM L., GAÏTI D., MHAMED A., "The Basis for an Adaptive IP QoS Management", *IEICE Transactions on Communications – Special Issue Internet Technology IV*, vol. E87-B, no. 4, p. 64-73, March 2004.

[CHR 01] CHRISTIANSEN M., JEFFAY K., OTT D., SMITH F.D., "Tuning RED for Web Traffic", *IEEE/ACM Transaction on Networking*, vol. 9, no. 3, p. 249-64, June 2001.

[FLO 93] FLOYD S., JACOBSON V., "Random Early Detection Gateways for Congestion Avoidance", *IEEE/ACM Transactions on Networking*, vol. 1, no. 4, p. 397-413, August 1993.

[GRO 02] GROSSMAN D., "New Terminology and Clarifications for Diffserv", *RFC* no. 3260, Internet Engineering Task Force, April 2002.

[HIC 98] HICKS M., KAKKAR P., MOORE J.T., GUNTER C.A., NETTLES S., "PLAN: A Packet Language for Active Networks", *Proc. of the Third ACM SIGPLAN International Conference on Functional Programming Languages*, ACM, p. 86-93, 1998.

[ITU 92] ITU-T, BISDN Reference Model, ITU-T Recommendation I.321, 1992.

[JAC 02] JACKSON A.W., STERBENZ J.P.G., CONDELL M.N., HAIN R.R., "Active Network Monitoring and Control: The SENCOMM Architecture and Implementation", *Proc. of DARPA active Networks Conference and Exposition (DANCE'02)*, 2002.

[JAI 91] JAIN R., *The Art of Computer Systems Performance Analysis*, John Wiley & Sons, 1991.

[KOH 01] KOHLER E., The Click Modular Router, PhD Thesis, Department of Electrical Engineering and Computer Science, Massachusetts Institute of Technology, February 2001.

[LE 03] LE L., AIKAT J., JEFFAY K., SMITH F.D., "The effects of active queue management on web performance", *Proc. of the 2003 Conference on Applications, Technologies, Architectures, and Protocols for Computer Communications*, ACM Press, p. 265-76, 2003.

[LIN 97] LIN D., MORRIS R., "Dynamics of Random Early Detection", *Proc. of ACM SIGCOMM'97*, Cannes, France, p. 127-37, October 1997.

[MER 02] MERGHEM L., GAÏTI D., "Behavioural Multi-agent Simulation of an active Telecommunication Network", *Stairs'2002*, Lyon, France, IOS Press, p. 217-26, July 2002.

[OPN 04] OPNET, OPNET Technologies Inc. Homepage, 2004, http://www.opnet.com.

[RAM 01] RAMAKRISHNAN K.G., FLOYD S., BLACK D.L., "The Addition of Explicit Congestion Notification (ECN) to IP", *RFC* no. 3168, Internet Engineering Task Force, September 2001.

[SCH 00] SCHWARTZ B., JACKSON A.W., STRAYER W.T., ZHOU W., ROCKWELL R.D., PARTRIDGE C., "Smart Packets: Applying Active Networks to Network Management", *ACM Transactions on Computer Systems*, vol. 18, no. 1, p. 67-88, February 2000.

[VEG 01] VEGESNA S., *IP Quality of Service*, Cisco Press, 2001.

[VUK 04] VUKADINOVIĆ V., TRAJKOVIĆ L., "RED with Dynamic Thresholds for Improved Fairness", *Proceedings of the 2004 ACM Symposium on Applied Computing*, ACM Press, p. 371-372, 2004.

[WAN 01] WANG Z., *Internet QoS: Architectures and Mechanisms for Quality of Service*, Morgan Kaufmann Publishers, 2001.

[YAN 04] YANG L., DANTU R., ANDERSON T., GOPAL R., "Forwarding and Control Element Separation (Forces) Framework", *RFC* no. 3746, Internet Engineering Task Force, April 2004.

Chapter 3

Software Agents for IP Management

3.1. Introduction

Telecommunications infrastructures are experiencing a very fast growth. The market liberalization has led to heterogenous and often incompatible telecommunications environments. In addition, technological advances and high throughput generalization have sparked a demand by users for more diversified services (in terms of functionality, quality, flexibility, etc.) and have increased the global load on networks. This network complication makes its management even more difficult. Current network management techniques are centralized and rely on hardware. Since network growth has caused flexibility and interoperability demands, the necessity for a distributed and "intelligent" control is gradually becoming obvious. For this reason, the last decade started seeing a connection between artificial intelligence, and especially distributed artificial intelligence (DAI) where the agent paradigm was created, and communications networks (CN). Cheikhrouhou *et al.* [CHE 98] cite the prevalence of object oriented programming (OOP) and the immaturity of programming by agents in explaining why only few agent paradigm applications exist in the network. However, in a competitive market searching for innovative solutions, the emergence of open standards in the multi-agent systems field leads us to consider solutions which integrate agents as viable alternatives, which are likely to benefit network operators, service providers and end-users.

Chapter written by Anneli LENICA.

The goal of this chapter is not to offer a complete state of the art on the subject, but rather to make a comparison between the two fields, which are traditionally apart, of multi-agent systems (MAS) and CNs through the agent paradigm applied to IP network management. We will identify current trends in network growth which would suggest the use of agents. These paths will be analyzed according to the characteristics of different network problems and the types of agent systems which could be associated with them. We will finish with an overview of the different research studies in this field.

Network management views discussed here will cover TCP/IP model transport layer functionalities. We will not discuss agents incorporated into applications such as “interface”, “user”, web agents, etc.

This chapter is divided into four sections. The first section is a reminder of current network problems. The second section presents the agent paradigm and MAS in an operational way. The third section reviews a certain number of studies conducted on IP networks. These examples are chosen as representative of the treated problems and types of implemented agent solutions. Finally, the last section will summarize the main points of this chapter and conclude with some MAS application perspectives in network management.

3.2. IP networks and their management

3.2.1. *IP networks*

The Internet network, a network of interconnected networks, is made up of a backbone, which is a high throughput fiber optic switched network connecting access networks nationally and internationally. For each access network (for the sake of simplicity, we will only call it network here) there is a domain managed by an administrator.

We show a network associated with a domain in a graph in Figure 3.1, where the tips correspond to the physical elements of the network (switches, dispatchers, routers and user stations) and the edges to communication channels between these elements. Among the routers there are edge and core routers. The edge routers can process flows traveling through them (marking, classification, admission control, etc.), whereas core routers only function as IP packet forwarders.

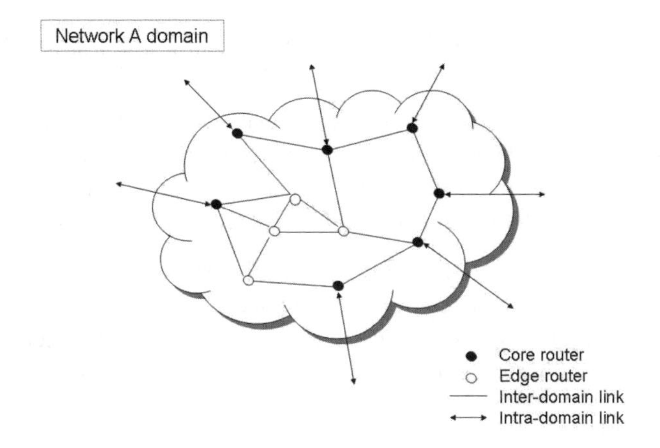

Figure 3.1. *An operator network and its connections with connected domains*

The size and complexity of a network can vary from a simple *intra muros* local network (LAN) to the millions of nodes that make up the Internet. As a network becomes more widespread, it verifies the following characteristics:

– *heterogenous:* networks are open systems. The diversity of the players and uses leads to a heterogenity of services and deployed network technologies. Internet is made up of a conglomerate of networks relying on different transport methods so obviously presents the highest degree of heterogenity. In this way, we observe a large variation of bandwidth capacity at different points in the network;

– *dynamic:* the network topology evolves (connecting stations to the network, faults, etc.). In addition, the sporadic nature of transported flows leads to high load variations. The state of connections between two hosts can vary from one millisecond to another;

– *distributed:* by definition, a network is a graph made up of physical elements linked together.

3.2.2. *IP network evolution and associated problems*

The structuring factors of network evolution are [WIL 00]:

– *deregulation of the telecommunications market.* It has led to a break in role distribution, economic models and the practices of network operators and service providers (their roles were actually only defined and differentiated in the early

1990s). In order to compete, the different economic players are competing for efficiency and search of added value through a diversified service offering;

– *rapid growth of technologies.* Combined with market deregulation, the diversification of deployed network technologies reinforces the heterogenity of CNs. The standardization efforts of committees such as ITU, ISO, ANSI, ATM Forum and IETF are not enough to correct the problem of technology interoperability;

– *flexibility requirements from users.* The diversification of the offering, the development of mobile technologies (mobile phones, laptop computers, etc.) and the generalization of high throughput concentrate the Internet service model towards a maximum information availability ("*any time, any place, any form information*") [MAG 96].

The problems resulting from these factors are many. Here are a few:

– *interoperability.* Among the different interoperability problems, we will focus on network integration. Currently several aspects of network integration are statically determined by contract (number and capacity of connections between two domains, pricing, etc.). This services interoperability management between different domains lacks flexibility and efficiency;

– *quality of service (QoS).* A large adaptability in service offer implies an adapted management of resources. Among these services, the multimedia applications in real-time and highly interactive (teleconferences, network games, etc.) are expanding rapidly. Contrary to applications called "elastic" such as file transfer (through FTP protocol) or email (with SMTP), this class has strict requirements in terms of resources, thus requiring services which are guaranteed, or at the very least predictive. However, the current Internet service model does not differentiate between flows and offers the same minimal QoS (*best effort*) regardless of the user or the application concerned. This explains today's demand for an evolution of the Internet model to a differentiation of services based on requirements (QoS);

– *automation.* Current networks are controlled at different organizational and functional levels by human agents. The automation of certain tasks such as switching or router control, or failure management, seems necessary, especially since networks are becoming more complex.

To respond to these different problems, research in the telecommunications field is increasingly driven by services. Consequently, network management is interchanging from the physical level to the application level in order to enable a better network integration, interoperability and flexibility.

3.2.3. *IP network management*

The notion of "network management" represents the implementation of a set of ways of assisting an operator in his network control and maintenance task.

The goal of a network operator is to make the best use of the current infrastructure. Two network management strategies can be considered: "all automatic" and manual configuration. Faced with increasingly difficult to operate and complex networks, the operator must find a compromise between these two options. The chosen strategy must be able to facilitate his task (by the automation of certain processes) and keep enough flexibility to dynamically adapt resources to requirements.

The needs for IP network management will vary depending on whether we are talking about an Internet access provider (IAP), a company network or other. IAPs, for example, are currently in fierce competition and because of this they strive to better control their resources: one may only be interested in information relative to general performances, whereas another may be concerned with the use of network resources or with a particular type of service. They dynamically wish to deploy different management strategies, which are adapted to their technical and commercial objectives (for example, controlling multimedia services duration as well as the quantity of information exchanged globally).

Current network management systems are based on centralized mechanisms. It is this way for SNMP (*simple network management protocol*), developed by IETF, and which has become the standard network management protocol for IP networks. With SNMP, management applications only access one manager, which questions SNMP "agents" present on the network's nodes to get information. All management applications are executed from a central server and the "agents" only serve as event listeners to the server and respond to its requests. The behavior of SNMP "agents" and the list of information that they monitor are completely static. It is therefore impossible to modify management policies dynamically and it is also impossible to distribute part of the management to network nodes/agents. The introduction of the "*proxy* agent" concept in SNMPv2 has established a hierarchical management that is less centralized but not really distributed.

Due to the complication of IP networks and their technical and economic challenges, the development of powerful tools for progressive and simplified management of IP networks has become a priority. The following sections will strive to show that the agent paradigm is an appropriate choice for the design of such tools.

The ISO model for network management is made up of five aspects: configurations, performances, anomalies, security and accounting information. In the context of what we are presenting, we will only discuss the first three points.

3.3. The multi-agent paradigm

The multi-agent paradigm is at the junction between artificial intelligence, distributed problem solution (DPS) and parallel programming. It provides a frame of reference for problems encouraging the design of flexible and adaptive architectures [MIN 99]. Thus, the inherent properties of agents facilitate data distribution and redundancy, parallel and asynchronous processing and semantic control of communicated information [BOU 01]. The agents can then contribute to the reliability and interoperability of systems.

3.3.1. *What is an agent?*

The definition of "agent" has become sensitive for two reasons [NWA 96]:

– its wide use in current language (realtor, etc.);

– its polysemy: its wide use in the field of information technology actually hides very diverse accepted variations of the word (confusion between mobile code and mobile agent, etc.).

In the absence of a definition that is unanimously accepted by the different "agent" user communities, we will limit ourselves to define and only use the term "agent" in this chapter. We will consider agents as software entities whose main property is autonomy. The property of autonomy means that the agent is considered able to act without human intervention or without the intervention of other systems: an agent controls its internal state and behavior. The notion of autonomy implies the notion of proactivity: the agents must be equipped with intention and not simply reacting to external stimuli [WOO 95]. This property of autonomy implies that the agents cannot be reduced to objects, expert systems or mobile code.

Associated with this fundamental property, qualities of interactivity (capacity to communicate with humans and other agents) and reactivity (according to external stimuli) are also shared by agents [JEN 98].

We can describe agents according to three dimensions:

– *their internal architecture:* Wooldrige and Jennings [WOO 95] classify agents depending on whether they are reactive, cognitive[1] or hybrid. Cognitive agents possess an internal representation of their environment and are capable of symbolic thinking and planning. Conversely, the behavior of a reactive agent is of stimulus/response type, with the stimulus resulting from a change in its environment state [FER 94]. Continuity between these two agent categories (on which the hybrid agents are located) should be considered;

– *their mobility:* an agent can be static or mobile (able to circulate in a network, for example). In this chapter we will not consider the capacity of a code (associated with data) to execute remotely as being enough to qualify it as a mobile agent. In general, we do not consider an isolated agent, but rather a population of agents interacting together. In this latter case we speak of an MAS. The affiliation of an agent to an MAS introduces another dimension to take into consideration in the comparison and distinction between (or among) studies;

– *the type of interaction within the MAS to which it belongs:* the agent can be cooperative or competitive. Cooperative agents will work together for the resolution of a common problem (we can then talk about collective rationality). Competitive agents will try to maximize their own usefulness function without considering the general interest (individual rationality). Again, this is not a rigid categorization: the behavior of an agent can have some "cooperative" and "competitive" traits.

As we will see, the most conclusive contribution of the agent paradigm resides in the agent's social aspects. A global behavior that is more complex than the sum of individual behaviors emerges from interactions between agents within an MAS. This is the reason we will speak of MAS instead of isolated agents for the remainder of this chapter.

The organization of agents in society has led to numerous studies on coordination techniques with the development of cooperation, negotiation and conflict management protocols among others, by analogy with human (requests for proposal, for example) or animal models. The MAS of ethological inspiration occupy a specific place in the context of this chapter. In fact, algorithms based on the function of certain insect colonies, specifically ants, have been applied successfully to many optimization problems, such as network problems. This new distributed optimization approach, which was introduced by [DOR 96], is known as "ant system" (AS) or even "ant colony optimization" (ACO). "Ant agents" distinguish themselves by an indirect communication mode through the environment, called stigmergy (for example, with trails of pheromones). The resulting coordination mode is based on positive or

1 They are also called "deliberative" agents.

negative retroaction. Besides the appearance of a relatively complex behavior from local interactions of low complexity [KAS 01], the other advantages of "ant colonies" are their extreme robustness and their massive scaling.

The examples in section 3.4.2 illustrate different types of possible interactions between agents in an MAS.

Finally, we quote Ferber [FER 89]: "an agent is a physical or abstract entity capable of acting on itself and on its environment with only a partial representation of this environment. It can communicate with other agents, it pursues an individual objective and its behavior is the consequence of observations, knowledge, competencies and interactions which it can have with the other agents in the environment".

This section does not attempt to be anything other than a simple introduction to the general MAS concepts. For a more detailed study, see: [JEN 95] for theory, languages and agent architectures, [NWA 96] for a discussion on agent applications, [HUH 98] for a review of AI and DAI techniques.

3.3.2. *When should MAS be used?*

There are significant works dedicated to MAS but the available information is divided between theory and methodology. Between the agent paradigm understanding stage and its application stage, the step of choosing (or not) an agent design has very little documentation available. Some authors such as Nwana adopt a critical approach of their field (see [NWA 99] for an update on the progress made between 1994 and 1999 in the agent field). Reading works from these authors makes the agent paradigm more operational by providing keys to understand the ins and outs of its use.

Jennings and Wooldridge [JEN 95] emphasize the fact that in order to be competitive, agent technology must enable one or the other: (i) to automate processes which could not be automated before, either because it was impossible (very rarely) or because the solutions using available technologies were considered too expensive (more likely); or (ii) to resolve problems already resolved in a better way (more natural, more effective, etc.). By taking these two criteria into consideration, we will observe common characteristics of classes of problems which can benefit from the multi-agent approach:

– *heterogenous:* the system's components are not compatible with each other. However, the inherent autonomy of agents presumes that the development of coordination and negotiation protocols will make them interact. A multi-agent modeling will explicitly take into account the interoperability aspects within a system;

– *dynamic:* the system's components are not necessarily known in advance or can change over time. Again, the autonomy property of agents occurs in their capacity to adapt to a changing environment. By definition an agent is able to perceive its environment and to act according to this perception;

– *distributed:* the system's components are located in different locations (spatial distribution). An MAS, as a society of agents, functions as a distributed system and therefore "naturally" models the inherently distributed problems. In the context of centralized systems, the use of mobile agents can "decentralize" processes. An example of an application is that of a code which must process data. Generally the data is brought into the code. However, if the cost of data transport is too high, it might be wise to bring the code to the data and the code can be encapsulated in a mobile agent;

– *agent metaphor:* the notion of agent or MAS can constitute the most natural representation of software functionality. An example is the simulation of a society of ants by an MAS where each agent represents an ant [FER 94];

– *pre-existing:* [JEN 95] and [NWA 99] mention the use of agents for the encapsulation of opaque systems. We will not discuss this point in detail.

Table 3.1 summarizes the information above.

System	Description	Contribution of agent paradigm
Heterogenous	Incompatible components	Ontologies, negotiation protocols, specialized agents
Dynamic	Temporal evolution	Flexibility, autonomy
Distributed	Spatial distribution	Robustness, parallelism scaling
"Agent" metaphor	Naturally represented by an agent or an MAS	Modeling simplicity
Pre-existing	Pre-existing component to reuse	Encapsulation

Table 3.1. *Characterization of potential applications of the agent paradigm*

During the design of a system, the technological choices depend on a large number of factors both technical and non-technical. This section is limited to the identification of problem classes for which an "agent" solution should be considered, without claiming that such a solution would necessarily be the best one in all cases.

Although MAS applications are numerous, this approach may not be suitable for all problems. It is the case with critical problems for which a provable solution is required. The situations presented above emphasize the adaptability and flexibility of agent-based solutions in comparison with other solutions. However, these properties are obtained to the detriment of code validation possibilities. In fact, these adaptability and flexibility properties are a direct consequence of the importance given to concepts of autonomy, cooperation and negotiation in the design of an MAS. Global system properties emerge from the initiatives and interactions of agents during program execution. The result is that the individual behavior of an agent can be specified during design, whereas the global behavior of an MAS cannot be set in advance. These systems need to use simulation in order for them to be studied since they cannot be understood analytically. Without being necessarily indeterminist and unpredictable as Jennings and Wooldridge [JEN 95] indicate, MAS still suffer from a lack of methodologies for proving "multi-agent" programs. The current prevailing methodology for MAS development is pragmatic and "intuitive" (of *trial and error* type). Although this approach may be suitable for prototyping, it does not work well with the development of operational solutions to industrial problems.

Agent technologies have nevertheless reached an adequate degree of maturity for them to be the object of a standardization process. The main players are OMG (Object Management Group) and FIPA (Foundation for Intelligent Physical Agents). With the MASIF (*mobile agent systems interoperability facilities specification*) project, OMG is working on the standardization of mobile agents in order to enable interoperability of mobile agent platforms via the use of Corba interfaces among others. FIPA propositions focus more on the high level aspects of interagent communication (languages, protocols, ontologies, etc.). The two initiatives complement each other and the future possibility of a synergy seems entirely possible. However, there is still a lot to do for MAS interoperability, especially with ontologies, as explained in [NWA 99].

3.4. MAS for IP network management

We have seen in the previous sections that multi-agent system characteristics correspond to the characteristics of networks (heterogenity, dynamism, data and resource distribution). The resulting possibility for modeling networks by MAS and the correspondence between network problem characteristics and the properties of the different agent types lead to the two network agent paradigm applications: network simulation by agents and network management by agents. The rest of the chapter explores these two channels based on a pragmatic and "network" oriented methodology. After becoming more familiar with the correspondence between some current case studies in network problems and the different agent categories in section 3.4.1, section 3.4.2 will discuss studies conducted in this field.

3.4.1. *MAS for specific network problems*

Reading the studies conducted on the resolution of network problems with agents causes two things to arise.

On the one hand, the term "agent" is sometimes used in an abusive manner as referred to at the beginning of this chapter. Too often, "mobile agents" are hard to separate from the mobile code, with often reduced "autonomy". Besides, the added value of "agent" models proposed in relation to "traditional" centralized models (client/server) or in relation to distributed models which are implemented according to other paradigms is not always emphasized enough.

On the other hand, the state of the art on the question of agents in networks such as [CHE 98, HAY 99, NWA 96, TSA 00] are presenting studies on the subject which are based on an "agent" perspective: the solutions are classified according to the category of agents used (starting from a solution describing the problems that it solves somehow). However, when faced with a given network problem, it is more useful to know which agent characteristics respond to the characteristics of the problem to solve. This section will focus on following this logic by identifying a few simple and concrete solutions in network management in order to suggest types of adapted agents. According to [NWA 99], the minimum decision criteria to justify an MAS-based solution must be the distribution of resources. From these criteria and the characteristics discussed in section 3.3.2, we can cite the following case studies (the list is not meant to be exhaustive, only indicative):

– *action depending on data distributed on several nodes.* Example: network control requires the retrieval of metrics over several nodes;

– *action which must be locally executed on one or several nodes.* Example: reconfigurations or updates which must be executed in several locations, simultaneously or not;

– *action depending on the coordination between several independent systems.* It may be the decision support systems for network management;

– *economy of resources.* Example: section 3.3.2 discusses the case where we need to execute a code over remote data. If the volume of data is larger than the code and the resources in terms of bandwidth are limited, it might be better to move the code to the data instead of moving the data to the code.

The constraints of the problem to process must obviously be considered during the design of an MAS. Clearly, reactive agents are more adapted to real-time execution. Another example: the symbolic reasoning of cognitive agents is an advantage when the MAS operation must be comprehensible or transparent for the human user [MIN 99].

3.4.2. *Existing applications*

This section provides a representative sample of proposed solutions using MAS in network management.

The aspects of network management covered here are: the development of topology maps, routing, congestion control, failure management and network monitoring, quality of service, continuity of service and simulation.

3.4.2.1. *Development of topology maps*

The goal of mobile agents of Minar *et al.* [MIN 99] is to determine the map of a wireless network. In order to do this, a set of agents covers the network. Each agent is equipped with a strategy of choice concerning the following related node to visit: i) it can simply choose a node at random, ii) it can choose a node which has never been visited or has been visited the least recently, iii) it can choose a node based on information from other agents. The authors test this model by varying the number of agents and the research strategies of agents.

The empirical study of the different multi-agent system parameters impact over global results is the main contribution to the study. The generalization of this type of experiment is meant to lead to a better understanding of the operation of MAS in general and it presents a real methodological point of interest. On the other hand, even if this aspect is not part of the object of the study, it is important to emphasize that numerous questions concerning the deployment of mobile agents still remain unanswered. The main drawback is most certainly the security linked to the principle of agent mobility [POS 00]: how should the host be protected from malicious agents? How should the agents be protected against future manipulations from hosts? How should the global operation of a mobile agent system be controlled?

3.4.2.2. *Routing*

The term "routing" covers two tasks: i) the distribution of information on topology and the state of the network (the concrete routing) and ii) the determination of routes which are actually used by data packets between <origin, destination> pairs (forwarding). The routing algorithms used are not adapted to current network growth [KAS 01]. Centralized algorithms present scaling problems, static algorithms are not reactive enough with respect to changes in the state of the network, and some dynamic and distributed algorithms show vacillating and unstable behaviors.

One of the most original approaches to the routing problem calls for ant colonies. It was introduced in [APP 94] and [SCO 96] (see below), then adapted to IP networks [SUB 97] and generalized to other protocols [CHE 99]. We present the principle here.

Schooderwoerd *et al.* [SCH 96] propose "ant-based control" (ABC), which is a detection algorithm for optimal routes based on an MAS represented by a community of "ants". This method is based on a network model that a group of artificial "ants" cover, with the responsibility of detecting optimal routes. The scattered "ants" move from one point of the network to another leaving a trail of simulated "pheromones" at each visited node. The quantity of "pheromones" is based on the degree of node congestion and on the distance covered from the source. The "ants" determine their route by choosing the related node with the highest concentration of "pheromones" at each hop. Since the "pheromone" is volatile, only the shortest routes are retained. These routes are calculated in advance and will be used for routing datagrams. An extension of ABC based on dynamic programming is proposed by [BON 98].

This approach by ant colonies is available in multiple ways. The best known algorithm with ABC is certainly AntNet [DIC 98] (which has also been the object of many variations), but we can also cite Adaptive-SDR [KAS 02] or [HEU 98] for packet switching networks. In circuit switching networks: White, Pagurek and Oppacher in particular have proposed RbA (*routing by ants*) [WHI 98A]. For a state of the art on the subject, see [KAS 01].

Certain solutions (AntNet, RbA, etc.) combine several types of agents and each is specialized in the resolution of an underlying problem (exploratory agents, RbA non-enablers, enablers).

Finally, except for the "ant" agents, we should mention another mobile agent application in routing: an extension of the routing algorithm with distance vectors (DVR) from mobile agents, proposed by Amin *et al.* [AMI 01]. The agents are used in this context to limit the increase of bandwidth usage linked to the exchanges between nodes from their respective routing table in case of network size increase.

3.4.2.3. *Congestion control*

The congestion appears when the incoming throughput is larger than the exiting throughput at router level. This leads to traffic disruption and a degradation of transmission quality.

The DWRED demonstrator [ROU 01] is an example of the application of an agent paradigm to the problem of congestion in networks. The agents are used in the extension of the RED (*random early detection*) router queue management algorithm. RED is an alternative to the router default behavior without quality of service guarantee. The queues of such a router operate by putting a certain volume of traffic on hold and by rejecting the "overflowing" packets. The processing of these packets is uneven and leads to retransmissions which may make the original congestion situation worse. Due to better control of the average queue occupation ratio, RED makes it possible to improve latency and to absorb traffic peaks in a better way, while ensuring a fairer packet processing. In addition, a variation of RED, i.e. WRED, enables a packet management with quality of service (QoS), i.e. a differentiation of the service based on packet priority. The goal of DWRED is to dynamically modify its parameters in order to optimize the network's behavior according to the traffic growth. This additional router control mechanism is implemented by cooperative agents. An agent corresponds to a router and each agent communicates with its neighbors (related routers). When a message is received, the agent reacts by modifying the queue parameters. In this way when an agent receives a message of <DEC, id> form, this will lead to the decrease of the minimum occupation threshold for that agent's queue for packet class "id".

The choice of reactive agents communicating by simple messages is certainly adapted to the problem considered. In fact, the agents are associated with core routers, which have restrictive performance constraints. It is therefore not possible in this case to use agents with complex reasoning and therefore costly in calculation time. In addition, transmitting messages during network congestion brings the risk of aggravating the situation with bandwidth overload as a result. The messages must remain as brief as possible. On the other hand, we can object that DWRED agents are autonomous since they “obey” their surrounding agents. It might have been more consistent with the principle of autonomy if the DWRED agents only informed their neighbors of their internal state and if the recipients of this information were left “free” to react to it. This encapsulation problem of the behavior greatly limits the implementation of different interaction policies. Other problems connected to the operation of open environment agents are not addressed, particularly the deployment of agents in routers and the behavior that should be adopted when the existing network does not enable the presence of an agent on each node.

3.4.2.4. *Network monitoring*

Section 3.2.3 introduced the SNMP protocol and showed its inadequacies to the requirements of network management players: what administrators are most lacking in is the lack of flexibility from current tools.

Although the study of the application of agents to management and network monitoring started early in the 1990s, works in this field are still not fully developed [BOU 01]. We can however mention the Intelligent Network project [ESF 96, ESF 98]. This project applies DAI techniques to the partial automation for dynamic processing of several alarms on the network and for notifications of received events by network management platforms. The developed system is able, for example, to filter unimportant events for a particular operator.

The proposed solution’s characteristic is to call for an “interface agent”. An “interface agent” is an agent whose role is to provide assistance to the end-user in a given application [MAE 93]. The role of this agent is primarily to learn, by observation, the behavior of a network monitoring operator and secondly to reproduce this behavior when the conditions leading to this knowledge are met.

This assistant was implemented and integrated in the (simplified) simulation program of a network management platform.

In order to improve interfacing with the user, the Intelligent Network agent is equipped with mental states, which must enable a better intelligibility of the agent’s behavior for humans. The agent has a learning capacity, which corresponds to the “traditional” definition of intelligence in computer programs. Even if the interface agent of the Intelligent Network operates alone, in close interaction with the end-user, its autonomy differentiates it from an expert system and classifies it among the agents. This autonomy is interesting in the sense that, whereas the expert system would limit itself to explaining the network’s failures, an agent can actively monitor the network, detect failures and launch alerts.

3.4.2.5. *QoS*

QoS is part of IP network growth problems as discussed in section 3.2.2.

[MEE 97] proposes a QoS architecture controlled by an MAS. The solution adopted presents the characteristic of including all the players (of a local network or WAN) in order to provide an end-to-end QoS. The architecture is broken down into functional layers (client, client system, network, server system, server), each one represented by a specific agent. This agent interacts with the other agents to ensure a satisfactory QoS. The operations that are executed are: the negotiation of QoS criteria, renegotiation, translation of the criteria from one layer to another, reservation of resources necessary for establishing the connection with the required quality, control of quality metrics and adaptation (operation of reallocating system resources based on global load). The interactions are "horizontal" and "vertical". Horizontally, they involve agents opposite each other (client and servers, client system and server system) in the processes of negotiation and renegotiation of QoS criteria. Vertically, the interactions issue connection requirements with a given QoS to adjacent layers. At each level, the agents are responsible for translating QoS criteria into an operational terminology for adjacent layers.

Resource optimization and procurement of QoS are two paradoxical objectives requiring negotiation procedures in order to lead to an acceptable compromise for all parties concerned. The underlying operations of reasoning, communication and decision are adequately modeled with the agent paradigm. To avoid having to call upon the user during negotiation or renegotiation steps, automatic learning techniques, ideally coming from the end user, could be applied to the level of user agents.

3.4.2.6. *Continuity of services*

Continuity of services is part of the interoperability themes presented in section 3.2.2. In the example described here, the problem we are dealing with is about bandwidth allocation in the case where transmitting and receiving flow hosts are not located in domains belonging to the same operators. The objective is to make operators "interwork" for end-to-end connections in a multi-operator network.

Network provider interworking (NPI) is an agent architecture explained in [CAL 99] and implemented in [CAL 00] for interdomain routing management with QoS. The interconnection of the different Internet network domains is modeled by a graph made up of peaks (the domains) and connected by edges (the group of interdomain links connecting two domains). The QoS architecture adopted is by reservation (resources necessary for the procurement of a given QoS are reserved along the route taken by the concerned flow). Each operator associated with a domain is represented by a *network provider agent* (NPA). NPA agents, which are distributed on the network, are responsible for two tasks: i) determine all the sections of the intradomain route and outgoing connections satisfying QoS criteria, and ii) negotiate with the other NPA agents in order to achieve an end-to-end connection. The constraints of the problem (QoS criteria and connectivity) are modeled in the form of a constraint satisfaction problem (CSP) which the NPA agents must resolve. The operator's graph is the only global knowledge required.

The advantage of negotiation automation between different players depends on the number of players present on the market [BOU 01]. In the very competitive telecommunications field, the NPI project appears to be completely relevant. NPI architecture makes up an abstract layer masking the characteristics of the different networks of the Internet. The NPA agent itself constitutes an abstraction of several specialized agents in the communication, negotiation, and application of CSP techniques, etc. The fulfillment of this abstraction must enable the centralization of the services provided by heterogenous networks (in this case: transmission of flows with a specific QoS). It implies a language which is common to all NPA agents. The use of techniques from FIPA (see section 3.3.2) and the definition of an adapted ontology made it possible to address the problems linked with agent interoperability. The negotiation process of NPA agents is about the connectivity of route sections which are locally represented. Another possible NPI architecture application would be connection pricing: NPA agents could be used to negotiate a route among available prices from the different domains, thus satisfying a compromise between the administrators' benefits and end-user needs. Automation by agents would be even more necessary in the case of dynamic pricing. The agents would not be cooperative anymore, but competitive.

3.4.2.7. *Network simulation*

De Meer *et al.* [MER 03] favor multi-agent modeling and simulation of telecommunications networks with the help of cognitive agents. Multi-agent modeling and simulation have been used for the empirical study of phenomena such as road traffic [ELH 00], the operation of ecosystems [DOR 01] or the impact of consumer choices on marketing strategy [BEN 02]. These phenomena and the network phenomena have in common the fact that they result from the interaction between a large number of autonomous players. The complexity of the underlying models makes their understanding, estimation and optimization difficult to achieve through traditional techniques. The simulation approach constitutes a first significant step in the study of these complex systems, before their formalization. It can take into account the dynamic, temporal and distributed aspects of modeled systems to provide an intuitive understanding of their behavior. The use of the agent paradigm within the simulation can be an additional advantage for the representation of autonomous and distributed players. In the IP network field, multi-agent simulation has received little attention until now, whether at the transport layer level or at the telecommunications market level (uniting operators, providers, regulating instances and clients), even though this chapter has shown that they constitute relevant MAS applications domains.

3.5. Perspectives and conclusion

This chapter has briefly shown the relevance of the agent paradigm in IP networks before evaluating a few examples of applications for managing these networks.

The use of agent-based solutions brings necessary dynamic flexibility for an "on the fly" implementation of specific and adaptable management strategies. This flexibility then can decentralize management to the network's nodes, thus increasing its reactivity.

Subsequently, the disintegration of management by agents reduces management complexity and uses AI/agent techniques to automate it.

However, the study of our examples emphasizes SMA implementation problems. We can say, as did Bouron [BOU 01], that the agent paradigm is only interesting when simpler solutions are not appropriate, as in the case of complex environments. Such an environment is characterized by a large number of players or components and large interaction possibilities (each component can interact with a large number of other components).

If the Internet is the best example of a complex environment, the network community is, however, still not sold on the use of agents in networks. Willmott and Calisti [WIL 00] explain this skepticism by the immaturity of agent technologies and by the ignorance of the IAD community about the network field. It does not seem necessary at this time to talk about a specific reluctance towards agents. The network's heterogenity and its complexity make any fundamental evolution long and difficult, regardless of the implemented technologies, as is shown by the resistance encountered in the deployment of the IPv6 protocol.

Despite these obstacles, the increasing number of publications in network conferences such as NetCon and Globecom, in addition to the continuing agent technology standardization efforts are indicative of an increased collaboration between the two communities.

Today the agent approach benefits from a technological and structural opportunity which should enable it to quickly make progress. This trend should rely on attempts to achieve "full-scale" systems instead of prototypes. This aspiration alone will enable the agent community to face the questions still being raised about the development, deployment and use of agent-based systems.

3.6. Bibliography

[AMI 01] AMIN K.A., MAYES J.T., MIKLER A.R., "Agent-Based Distance Vector Routing", *MATA*, 2001.

[APP 94] APPLEBY S., STEWARD S., "Mobile Software Agents for Control in Telecommunications Networks", *British Telecom Technology Journal*, 12(2), 1994.

[BEN 02] BEN SAID L., BOURON T., DROGOUL A., "Agent-based interactions analysis of consumer behaviour", *AAMAS'2002*, Bologna, Italy, 2002.

[BON 98] BONABEAU E., HENAUX F., GUERIN S., SNYERS D., KUNTZ P., THERAULATZ G., "Routing in telecommunications networks with "smart" ant-like agents", *Intelligent Agents for Telecommunications Applications '98*, 1998.

[BOU 01] BOURON T., "Application des systèmes multi-agents dans les télécommunications", *Principes et architecture des systèmes multi-agents*, Traité IC2, Hermes, 2001.

[CAL 99] CALISTI M., FREI C., FALTINGS B., "A distributed approach for QoS-based multi-domain routing", *AiDIN'99, AAAI-Workshop on Artificial Intelligence for Distributed Information Networking*, 1999.

[CAL 00] CALISTI M., FALTINGS B., "Agent-Based Negotiations for Multi-Provider Interactions", *Proceedings of ASA 2000, 2nd International Symposium on Agents Systems and Applications,* Zurich, Switzerland, 2000.

[CHE 98] CHEIKHROUHOU M.M., CONTI P., LABETOULLE J., "Intelligent Agents in Network Management, a State-of-the-Art", *Networking and Information Systems*, 1(1), p. 9-38, 1998.

[CHE 99] CHEN J., DRUSCHEL P., SUBRAMANIAN D., "A new approach to routing using dynamic metrics", *Proceedings of INFOCOM99*, 1999.

[DIC 98] DI CARO G., DORIGO M., "AntNet: Distributed Stigmergetic Control for Communications Networks", *Journal of Artificial Intelligence Research*, vol. 9, p. 317-365, 1998.

[DOR 01] DORAN J., "Agent-based Modelling of Ecosystems for Sustainable Resource Management", *EASSS'01*, Prague, Czech Republic, 2001.

[DOR 97] DORIGO M., GAMBARDELLA L.M., "Ant Colony: A Cooperative Learning Approach to the Traveling Salesman Problem", *IEEE Transactions on Evolutionary Computation*, 1(1), p. 53-66, April 1997.

[ELH 00] EL HADOUAJ S., DROGOUL A., ESPIE S., "How to Combine Reactivity and Anticipation: The Case of Conflict Resolution in Simulated Road Traffic", *MABS'2000 workshop*, Boston, USA, 2000.

[ESF 96] ESFANDIARI B., DEFLANDRE G., QUINQUETON J., DONY C., "Agent-oriented techniques for network supervision", *Annals of Telecommunications*, 51(9-10), p. 521-529, 1996.

[ESF 98] ESFANDIARI B., DEFLANDRE G., QUINQUETON J., "An interface agent for network supervision", in Albayrak S. (ed.), *IATA 1996 (Intelligent Agents for Telecom Applications)*, IOS Press Publisher, p. 21-28, 1998.

[FER 89] FERBER J., Objets et agents: une étude des structures de représentation et de communication en IA, Thesis, University of Paris VI, June 1989.

[FER 94] FERBER J., "Simulating with Reactive Agents", Hillebrand, E. and Stender, J. (ed.), *Many Agent Simulation and Artificial Life*, IOS Press, Amsterdam, p. 8-28, 1994.

[HAY 99] HAYZELDEN A.L.G., BIGHAM J., "Agent Technology in Communications System: An Overview", *The Knowledge Engineering Review*, 1999.

[HEU 98] HEUSSE M., SNYERS D., GUERIN S., KUNTZ P., "Adaptive agent-driven routing and load balancing in communication network", *Proceedings of ANTS'98, First International Workshop on Ant Colony Optimization,* Brussels, October 1998.

[HUH 98] HUHNS M.N., SINGH M.P., *Readings in Agents*, Morgan Kaufman, 1998.

[JEN 95] JENNINGS N.R., WOOLDRIDGE M., "Applying agent technology", *Applied Artificial Intelligence*, (9)4, p. 351-361, 1995.

[JEN 98] JENNINGS N.R., SYCARA K., WOOLDRIDGE M., "A Roadmap of Agent Research and Development", *Journal of Autonomous Agents and Multi-agent Systems*, (1)1, 1998.

[KAS 01] KASSABALIDIS I., EL-SHARKAWI M.A., MAKS II R.J., ARABSHAHI P., GRAY A.A. "Swarm Intelligence for Routing in Communication Networks", *IEEE Globecom 20001*, San Antonio, Texas, 2001.

[KAS 02] KASSABALIDIS I., EL-SHARKAWI M.A., MAKS II R.J., ARABSHAHI P., GRAY A.A., "Adaptive-SDR: Adaptive Swarm-based Distributed Routing", *Proc. IEEE World Congress on Computational Intelligence*, p. 12-17, Hawaii, 2002.

[MAE 93] MAES P., KOZIEROK R., "Learning interface agents", *Proceedings of the 11th Nat Conf on Artificial Intelligence AAAI*, MIT Press/AAAI Press, 1993.

[MAG 96] MAGEDANZ T., ROTHERMEL K., KRAUSE S., "Intelligent Agents: An Emerging Technology for Next Generation Telecommunications?", *INFOCOM*, San Francisco, 1996.

[MEE 97] DE MEER H., PULIAFITO A., TOMARCHIO O., "An agent-based framework for QoS management", *4th International Conference on Analytical and Numerical Modeling Tech.*, Singapore, 1997.

[MEE 98] DE MEER H., PULIAFITO A., TOMARCHIO O., "Management of QoS with Software Agents", *Cybernetics and Systems: An International Journal*, 29(5), 1998.

[MER 03] MERGHEM L., GAÏTI D., PUJOLLE G., "On using multi-agent systems in end to end adaptive monitoring", *LNCS 2839*, p. 422-435, Kluwer Academic Publishers, 2003.

[MIN 99] MINAR N., KRAMER K.H., MAES, P., "Cooperating Mobile Agents for Mapping Networks", *Proceedings of the First Hungarian National Conference on Agent Based Computation*, 1999.

[NWA 96] NWANA HYACINTH S., "Software agents: An Overview", *The Knowledge Engineering Review*, 11(3), p. 205-244, October/November 1996.

[NWA 99] NWANA HYACINTH S., NDUMU D.T., "A Perspective on Software Agents Research", *The Knowledge Engineering Review*, 14(2), p. 1-18, 1999.

[POS 00] POSEGGA J., KARJOTH G., "Mobile Agents and Telco's Nightmares", *Annals of telecommunications*, 55(7-8), p. 29-41, 2000.

[ROU 01] ROUHANA N., HORLAIT E., "Dynamic Congestion Avoidance using Multi-Agent Systems", *MATA'01*, 2001.

[SCH 96] SCHOODERWOERD R., HOLLAND O., BRUTEN J., ROTHKRANTZ L., "Ant-based load balancing in telecommunications networks", *Adaptive Behavior*, 5(2), p. 169-207, MIT Press, 1996.

[SCH 97] SCHOODERWOERD R., HOLLAND O., BRUTEN J., "Ant-like Agents for load balancing in Telecommunications Networks", *Proceedings of Agents'97*, Marina del Rey, p. 209-216, ACM Press, 1997.

[SUB 97] SUBRAMANIAN D., DRUSCHEL P., CHEN J., "Ants and Reinforcement Learning: A Case Study in Routing in Dynamic Networks", *Proceedings of IJCAI'97*, p. 832-838, 1997.

[TSA 00] TSATSOULIS C., SOH L.K., "Intelligent Agents in Telecommunication Networks", in W. PEDRYCZ, A.V. VASILAKOS (ed.), *Computational Intelligence in Telecommunications Networks*, CRC Press, 2000.

[WHI 98a] WHITE T., BIESZCZAD A., PAGUREK B., "Distributed Fault Location in Networks Using Mobile Agents", in ALBAYRAK S. and GARIJO F.J. (ed.), *Proceedings of the Second International Workshop on Intelligent Agents for Telecommunication (IATA'98)*, 1437, Springer-Verlag, Heidelberg, Germany, 1998.

[WHI 98b] WHITE T., PAGUREK B., OPPACHER F., "Connection Management using Adaptive Mobile Agents", *Proceedings of 1998 International Conference on Parallel and Distributed Processing Techniques and Applications*, 1998.

[WIL 00] WILLMOTT S., CALISTI M., "An Agent Future for Network Control?", *Swiss Journal of Computer Science*, p. 25-32, January 2000.

[WOO 95] WOOLDRIGE M., JENNINGS N.R., "Intelligent Agents: Theory and Practice", *The Knowledge Engineering Review*, 10(2), p. 115-152, 1995.

Chapter 4

The Use of Agents in Policy-based Management

4.1. Introduction

The IETF (Internet Engineering Task Force) specified a policy-based control architecture to manage IP networks with quality of service guarantees. These policies define a desired behavior for network components in order to respond to demands from the different applications.

The appearance of this policy-based management concept comes from the need to simplify router configuration with an automatic mechanism. This automation has become necessary since the network continuously increases in size. Several fields are interested in this policy-based management and the most advanced focus on quality of service.

For large networks with frequent changes in operation instructions, policy-based management provides an attractive solution since it dynamically translates the objectives of the company into applicable network configurations over a range of components. Compared to other management technologies, policy-based management will adapt more quickly to changes in management demands after deployment, or to modifications of the environment with a dynamic policy rule update.

Chapter written by Francine KRIEF.

Policy-based management is also called policy-based control when the aspects linked to the control are favored. With policy-based control, the network operator can intervene at a higher level than at component level without worrying about implementing control parameters. Control and management are actually integrated in one information model called PIB (*policy information base*) [HAM 00]. Policies installed by network management using the PDP-PEP model will dictate network control behavior. We use the term policy-based management here whether we discuss policy-based control or management.

The introduction of intelligence in the network through multi-agent systems (MAS) makes it possible to build flexible systems with complex and sophisticated behaviors through the combination of highly modular components. The intelligent components (agents) and their interaction capabilities form an MAS.

In general, an agent is considered as a software element responsible for the partial execution of a process. It contains a certain level of intelligence, ranging from simple predefined rules to self-taught artificial intelligence machines. It typically acts in the name of a user or a process making it possible to automate a task. Generally, the term "intelligent agent" will have a wide range of meaning from adaptive user interfaces, also known as interface agents, to intelligent processes cooperating between each other for the execution of a common task [BRI 01].

These agents are able to bring their expertise to control network infrastructures. Their properties of autonomy in particular and of adaptation and communication respond well to network management problems. By using agents we can automate control and management processes and better adapt services to user requirements [AGO 00]. The network then becomes a "smart network", which means that it is able to adapt to a new situation, to control the sensitive states and to manage services in unforeseen conditions [PUJ 04].

Studies on agent introduction in policy-based management have mainly focused on dynamic service negotiation, procurement and control. In highly dynamic environments, such as wireless networks, mobile agents are widely used. A mobile agent is usually defined as an independent program acting on behalf of a user or another entity and is able to move from one network to another [FER 99, KRA 96, BRE 98].

This chapter is organized as follows: section 4.2 presents policy-based management, its advantages compared to traditional management and the major challenges. It also discusses agent characteristics and their contribution in network management. Section 4.3 gives an overview of agent approaches in dynamic service procurement and control in wired and/or wireless environments. Section 4.4 is dedicated to the use of intelligent agents in the dynamic negotiation of intra and interdomain services for mobile or fixed users. Section 4.5 discusses the contribution of agents in policy-based management of emerging services. Section 4.6 concludes this chapter by focusing on the ongoing problems and the corresponding research fields.

4.2. Policy-based management

Policy-based management was defined by the IETF within the IntServ model [YAV 00]. The proposal was to use a policy-based environment to manage the admission control of network resources reservations. However, this method is independent of any specific service model and can be applied to an environment of differentiated services [BLA 98], of IP-Sec security and even to any network environment [MAR 96] including TMN (telecommunications management network) [HAM 00].

Compared to traditional network management approaches such as TMN or TINA-C (Telecommunication Information Networking Architecture Consortium), policy-based management focuses on users and applications instead of on equipment and interfaces [YAN 02].

The objective of policy-based management, as defined by IETF, is to propose an infrastructure which provides a certain level of abstraction and enables a network behavior that is flexible towards the different events which can happen during its management, through the notion of policy.

4.2.1. *The policies*

Policies can be defined as a set of rules capable of managing and controlling access to network resources. They also help network administrators or service providers to control network component behavior based on different criteria such as user identity or application type, traffic requested, etc.

In general, policy rules are presented in the following form:

IF condition_activation_policy THEN action_policy

The condition indicates when the policy can be activated. For example, monitoring network status, node or type of traffic events entering the network can launch a policy action which will result in network resource provisioning.

The idea is to centralize policies in a database and to guarantee their application by distribution mechanisms on what we call an administrative domain which represents the policy range defined by the administrator [BAR 01].

A policy can be defined at several levels. The highest level is the *business* level. This policy must then be translated into a network language and then into a low level language that hardware will understand.

4.2.2. *Information model*

The IETF, in conjunction with DMTF[1] (distributed management task force), is working on information models linked to different language levels. The goal of this information model is to define a general model which can adapt to the different network management and control domains independently of the hardware type. The core of the policy environment information model called PCIM (*policy core information model*) [MOO 01] is an extension of the CIM (*core information model*) model of DMTF. The network is considered as a state machine where policies are meant to control status changes. This model must be able to identify and model current states and define possible transitions from rules describing the policies. It defines the roles, priorities and execution sequence, but is still abstract for the objects.

Two extension levels have been defined for QoS (quality of service) (Figure 4.1):

– QPIM model (*QoS policy information model*) [SNI 01] defines the precise actions to execute on packets. It integrates specific QoS notions in order to create formal representations of abstract policies. For this, the model defines policy actions and a traffic model to specify the management of a request or the arrival of a flow;

– QDDIM model (QoS *device datapath information model*) defines actions to take on equipment, i.e. on their configuration [MOO 02].

---

1 http://www.dmtf.org.

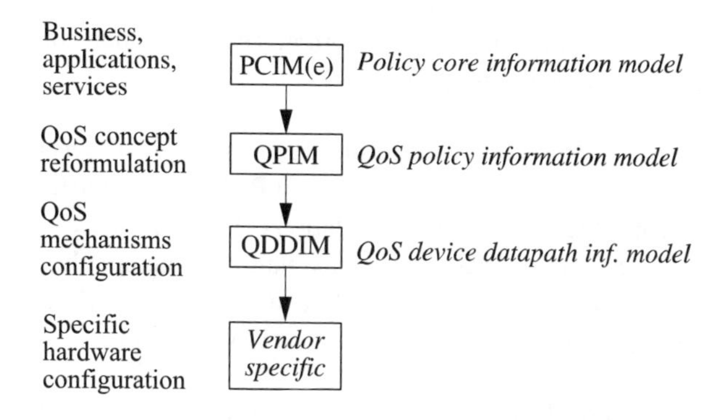

Figure 4.1. *Information model*

4.2.3. *Architecture*

The architecture defines a centralized model for managing, storing policies, decision-making and distributing configuration parameters to routers. This manager/agent type architecture is described in Figure 4.2.

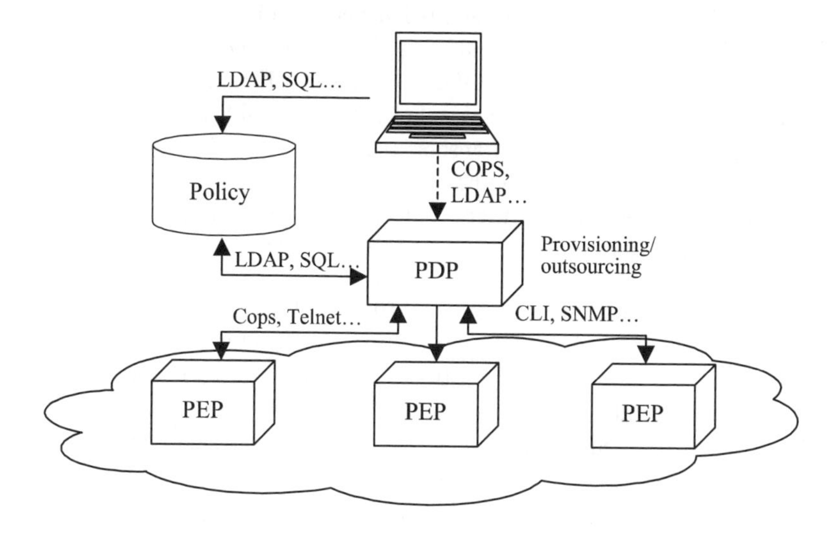

Figure 4.2. *Policy-based management architecture*

PDP (*policy decision point*) must determine the configuration to implement and the resources to use to satisfy the demand. Its main functions include the determination of policy rules to apply to the different PEPs (*policy enforcement point*), their conversion into an adapted format (PIB or other solution) and the guarantee of their adequate distribution.

A PEP corresponds to resources offering different types of services, which are configured to execute the policies determined by PDP. The main functions of PEP consist of connecting the external representations (PIB, MIB, etc.) of the internal configuration of the network hardware and maintaining a compatibility between the applied policies. When the PEP is located in an edge network element (access router), it is also responsible for uploading requests to the PDP.

The architecture model does not need any specific communication protocol or access method to a server storing policies. However, the LDAPv3 protocol seems to be the most accepted solution to communicate with a policy directory and the COPS protocol was defined to transport policies.

4.2.4. *COPS protocol*

COPS (*common open policy service*) [BOY 00] is a simple query/response protocol based on the TCP protocol and is independent of the information model. It was proposed by the IETF's RAP (*resource allocation protocol*) group to transport policies between PDP and PEP [YAV 00]. COPS is a flexible protocol because it can be applied to several policy domains. For example, COPS-RSVP is an extension of the COPS protocol. Its objects transport policies for RSVP messages admission control. COPS-PR [CHA 01] is another extension of the COPS protocol. Its objects transport policies to configure DiffServ routers. COPS-IP-TE is another extension of the COPS protocol and its objects transport policies for traffic engineering.

COPS-SLS [NGU 01] is an extension of the COPS protocol for dynamic SLS (*service level specification*) negotiation. An SLS [GOD 00] is a set of parameters and their values defining the service offered for data flow. To negotiate a level of service with his provider, a client sends the appropriate SLS to his ISP (*Internet service provider*). The ISP can accept or reject the SLS or propose another level of service to his client. Once an agreement is reached between the two parties, a contract is established and the user's data can travel through the ISP with the negotiated level of service. Policies reflect the ISP's negotiation strategy and enable the PDP to know how to respond to an SLS request. Figure 4.3 illustrates the model introduced by COPS-SLS.

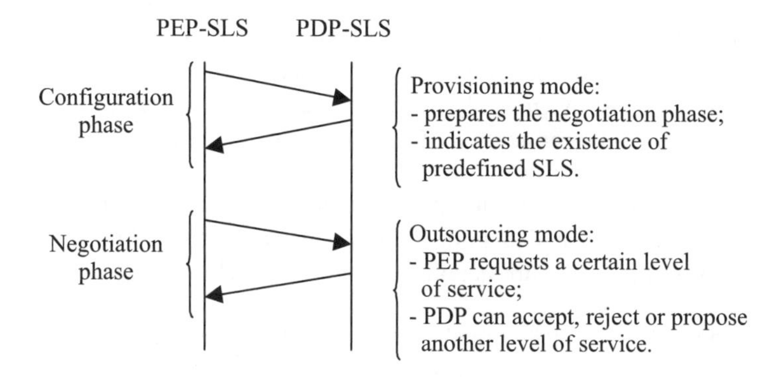

Figure 4.3. *COPS-SLS protocol*

4.2.5. *Advantages and challenges*

The main advantage of policy-based management is its ability to automate the establishment of company objectives over a wide range of network hardware [BOU 95]. The administrator can then interact with the network by providing high level abstract policies, i.e. independent from the hardware.

The automatic translation process will hide the complexity of network configuration construction. Network policies come from high level policies which facilitates the connection between company objectives and network configurations. In addition, such automation makes it possible to provide a more coherent and integrated representation of company objectives.

The dynamic modification of policies would provide a flexible way to control the network's behavior when the environment changes. This would also provide more freedom to users and service providers for a dynamic description of their policy requirements. Furthermore, a coherent operation without human intervention would be possible.

For Hamada, Czezowski and Chujo, policy-based management provides a richer context for more intelligent IP network management because each policy rule can be seen as an element of an intelligent distributed information system whose expert knowledge is strictly focused on the management and operation of an IP network [HAM 00].

The significant challenge in automating the management process is the automatic translation of a service contract (SLS – *service level specification*) into policy rules, on the one hand, and of policy rules received into policies that will ultimately be understood by the hardware on the other hand. This process is even more difficult to implement in a multidomain environment because administrators (intradomain service providers) must dynamically establish interdomain service contracts for the delivery of an end-to-end service and SLS compliance assurance.

The introduction or modification of new policies must not disrupt the system. It is therefore necessary to validate policies before they are deployed in the network in order to avoid conflict with current policies, for example [LUP 97].

The dynamic and automated modification of policy rules requires the definition and implementation of new mechanisms.

Several studies suggest solutions for these different problems, often using the agent technology.

4.2.6. *The agents and their interest in network management*

An intelligent agent is an independent entity, which is able to execute complex actions and solve management problems autonomously. In order for this to work, it must receive high level objectives with, for example, the help of policy rules [FIP 01, SER 03].

Wooldridge and Jennings [WOO 95] have defined three types of intelligent agent architecture: deliberative, reactive and hybrid agents. The deliberative agents have the capability, due to an explicit representation of the world, to generate intelligent actions. Reactive agents, on the other hand, require no representation of the world. They generate behaviors strictly based on environment observations. In reality, they are more reactive than deliberative agents because of their lack of complex symbolic reasoning mechanism. They could successfully be used for traffic monitoring, congestion control and admission control because these management functions require quick actions [BOU 95]. Hybrid agents are composed of both deliberative and reactive agents. A hybrid agent could contain a model of the symbolic world, develop plans and make decisions in the same way as a deliberative agent. However, it is also able to react to events happening in the environment without a need for complex reasoning. The hybrid agent seems well adapted for fault diagnosis [CHI 98].

Intelligent agents present interesting characteristics such as autonomy, sociability, reactivity, proactivity, learning and mobility. In this latter case, the agents transport their code, their execution states and maybe even their data

[BAL 98]. Most of the current studies on mobile agents focus on the reduction of network traffic and asynchronous interaction [KAR 03].

The introduction of agents in policy-based management will enable policy adaptation to environment changes. This has many advantages such as:

– a higher degree of freedom, for example, by offering users the possibility to modify their needs dynamically during communication;

– the introduction of policy-based management in highly dynamic environments such as wireless networks;

– a modification in network behavior in order to prevent any service contract intrusion or violation risk.

Agent technology will also be used to automate the negotiation process of service contracts between different administrative domains. The complexity of the problem resolution is left to the agent which is guided by high level policies. In this context, mobile agents are well adapted to wireless environments or when a user is mobile because they are not sensitive to connection breaks and they offer a good bandwidth saving. In this way, a wireless user sends his proposition via a negotiation agent and disconnects. When it connects again, the mobile agent moves to its local environment and informs the user of the negotiated parameters. In all cases, the use of mobile agents is justified when the interaction volume and complexity make the traditional client/server implementation difficult.

Finally, intelligent agents offer quick, flexible and automatic deployment.

We will discuss the contribution of agents in three main categories of policy-based management:

– provisioning and service control;

– service contract negotiation;

– management of emerging services.

4.3. Provisioning and service control

Policy-based management and agent technology are approaches that work well together. On the one hand, the agents can implement policy-based management systems in a flexible way. On the other hand, agents have an intelligence which can be guided by the establishment of policies.

We present architectures which have been proposed to enable the dynamic service control and provisioning. The first two examples use intelligent agents for the dynamic control and provisioning of QoS, in wired and wireless environments. The two following examples use mobile agents: one to manage QoS and the other to manage mobile users. The last example uses BDI agents for the detection of intrusions.

4.3.1. *Dynamic QoS provisioning in wired networks*

Project I3 (Internet third generation) is one of the first projects to propose an architecture pairing policy-based management and agent technology for the management of future IP networks.

4.3.1.1. *Project I3*

Project I3 is a cooperative research action proposed by INRIA whose aim is to reflect on long-term implementation needs to manage future IP networks. In this context, agents have been added to the core of a policy-controlled network in order to facilitate its management and to respond to QoS requirements from clients [BOU 01].

A functional architecture was initially proposed to ensure end-to-end quality of service in a policy-managed IP network [CHA 01]. This architecture is made up of four main parts:

– contract management is responsible for all the activities related to the contract. It groups the Subscription contract and Invocation contract functions;

– network management is responsible for traffic management on the network's physical resources as well as for network configuration according to the contract. It includes two functions related to routing and resource reservation respectively;

– policy management is responsible for storing management policies and also offers a tool for managing these policies;

– monitoring evaluates the current network state. It analyzes network behavior during a period of time and reports QoS parameter results concerning the network and the service contract to the policy-based management system. Monitoring, in this case, is not only used for diagnosis but it also makes the network reactive and perhaps even proactive.

Agents have been placed at the network's core to carry out the functionalities described previously. These agents are organized in an MAS for which an activation structure has been defined. This activation structure is made up of reactive agents organized in three classes [CAR 00]:

– interface agent collects information on its environment;

– activation agent receives the information collected, applies a filter to the data and transmits it to the motricity agent;

– motricity agent commands the system components based on the deliberative agent's decisions.

The MADKit[2] platform has been used to implement this environment. This platform is based on the AALAADIN [FER 98] model which combines the different agents by groups and gives them roles.

In this context, two agent models have been proposed: the PEP-router and the PDP models.

The PEP-router agent model, shown in Figure 4.4, contains these main elements:

– *stream agent* is defined to ensure the flow management for the duration of the communication. The *stream agent*'s main goal is to provide information about demands in terms of QoS;

– *manager agent* must combine information about aggregated flow requirements to send to the *information agent* from the "Managers" group;

– *information agent* possesses a global vision of all client requirements. It transmits a service requirement to the PDP which then decides what policies to apply in order to satisfy the demands;

– *mon agent* is an activation agent. It communicates with interface agents, i.e. the *IPPM agents*. These agents are responsible for collecting QoS indicators. If it detects exceeding contract data, *mon agent* reports it to its PDP counterpart;

– *net agent* is a motricity agent that will apply the decision coming from the PDP. In order to do this it will call the *allocation agent*.

2 Multi-agent development kit.

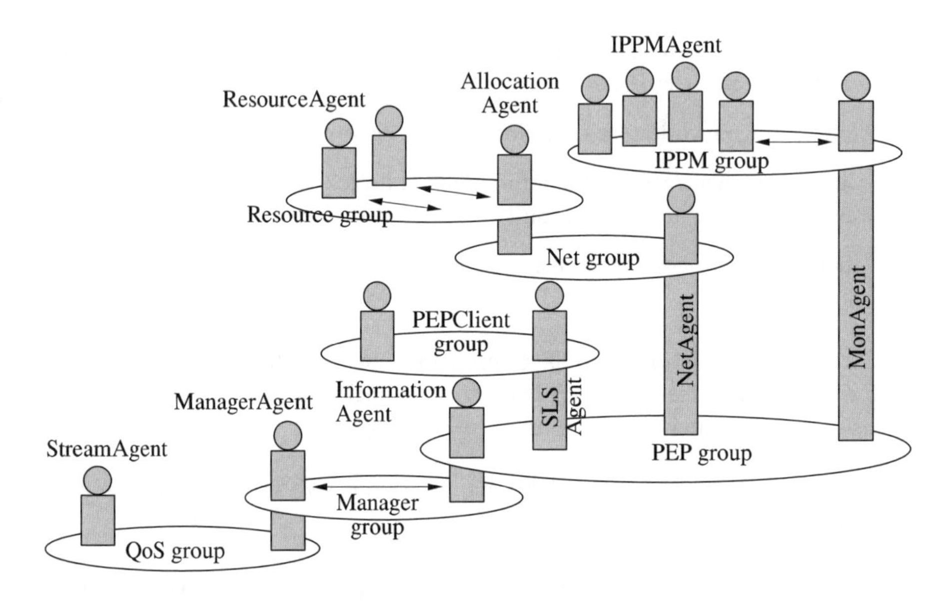

Figure 4.4. *PEP-router agent model*

4.3.1.2. *PDP agent model*

The PDP agent model is based on the functional architecture described earlier. It contains the following groups:

– *PDP client group* carries out admission control. It is also responsible for clients demanding new service contracts;

– *PDP monitor group* makes sure that the service contract is respected. It is also responsible for monitoring the network as well as for its resizing;

– *PDP net group* contains the agents responsible for routing and resource allocations.

The introduction of agents in this architecture provides numerous advantages such as the possibility to react in the case of SLA violation risk and to adapt the network's behavior to changes in service requirements from clients during communication via the COPS-SLS protocol, for example.

4.3.2. *Dynamic QoS provisioning in wireless networks*

Although several studies have focused on policy-based management in wired networks [GAY 02, FLE 02], few studies focus on policy-based management in wireless networks. The highly dynamic nature of wireless networks, characterized by frequent changes in resource availability, makes it impossible for current policy-based environments to be directly applied to them.

Intelligent agents are well adapted to highly dynamic environments. They are autonomous and proactive, i.e. they self-react to environment changes which present opportunities. They are also able to improve their performance by reusing past experience.

Samaan and Karmouch propose an adaptive context for QoS management in wireless networks by using policy-based management and agent technology [SAM 03].

The architecture proposed is a multi-agent multilayer system. In the prediction layer, QoS prediction agents use information from the user such as profile, location and characteristics of the station to predict possible changes which may affect the level of QoS provided. These changes are then transmitted to a set of adaptation agents located in a second layer, which will dynamically modify policies to apply to resources in order to respond to user needs. To help adaptation agents in reaching their decision, the monitoring agents, which are located in a third layer, provide them with information on the actual QoS. Below, we explain these three layers in more detail.

4.3.3. *Prediction layer*

It contains four types of agents, each completing a specific task:

– *user agent* helps the user express his needs in terms of policy. In order to do this, it utilizes a user interface. It can also learn the user's needs by analyzing frequently used activities;

– *location agent* predicts the user's location;

– *application agent* identifies QoS application constraints;

– *prediction agent* is responsible for collecting the different requirements, for translating them into QoS goals to reach, for possible events to apply on network resources and for constraints to follow. It then transmits this information to the QoS adaptation agent.

4.3.4. *Adaptation layer*

It is responsible for adapting policies to user requirements. This adaptation can be done in three ways [LYM 02]:

– by dynamically changing QoS policy parameters to specify new attribute values;

– by making a policy taken in a predefined set of QoS policies available or not;

– by creating new policies.

The third approach is the one retained here because of its highly dynamic environment. The *adaptation agent* first indicates its objectives (values of QoS parameters requested by the prediction agent), its constraints and the different possible actions, each with its own benefit function indicating its success rate. It can negotiate with other adaptation agents in order to exchange objectives, constraints, actions and even modify the values based on benefit functions. The adaptation agent then selects one or more actions for reaching its objectives. The policy (or policies) associated with these actions are then deployed in the network. The degree of success of these policies is then evaluated based on QoS measures coming from the *monitoring agent*. In order to achieve this, a re-evaluation module will modify the benefit functions associated with actions to encourage or discourage their future use.

4.3.5. *Monitoring layer*

The monitoring layer is responsible for measuring QoS in the network. It is made up of *monitoring agents* which cooperate to monitor the different QoS parameters. A monitoring agent can interact with others to combine the required QoS measures, which are then sent back to the adaptation agent. This agent must tell the monitoring agent which parameters to measure in order to evaluate applied network policies.

With the proposed approach, we can group policies in real-time according to a series of imposed constraints and objectives to satisfy. It offers more flexibility to users and applications by enabling the dynamic change of QoS parameters while maintaining a steady QoS flow. Predicting changes which may affect QoS in highly dynamic environments such as wireless networks seems vital in order for these environments to adapt in an intelligent way and to ensure the required QoS. In addition, the approach retained will make it possible to improve selected strategies through a learning process.

4.3.6. *Mobile agents for policy-based QoS provisioning*

We present here the work accomplished within the European MANTRIP[3] project on applications for QoS configuration.

4.3.6.1. *IST[4] MANTRIP European project (IST-10921)*

The main goal of project MANTRIP (*management testing and reconfiguration of IP based networks using mobile software agents*) is to provide IP network management applications by using mobile agents.

The MANTRIP network management system is based on policy-based management and contains four layers [YAN 02]:

– *application layer* includes user applications and policy-based management tools;

– *service layer* contains MANTRIP management services. PDP is located at this level;

– *adaptation layer* connects the service and resource layers. It contains the PEPs;

– *resource layer* contains the resources to manage.

All layers, with the exception of the lowest layer containing the elements to control, were developed with *grasshopper* mobile agent platform. All components described in the architecture are either fixed or mobile agents depending on their task.

Mobile agents, controlled by policies, self-migrate to specific PEPs to apply the policy. PEPs are established as fixed agents communicating with mobile agents from the service layer. Mobile agents are also responsible for transporting policies throughout the different domains.

This architecture has been tested within interdomain IP VPN[5] provisioning. These tests have shown the advantage of an architecture which combines mobile agent technology with policy-based management.

This approach enables quick and automatic deployment of new services, increasing load capacity and cost reduction in the network and service management. The mobile agent decreases communications between PDP and PEP. It also

3 MANTRIP project website: http//www.solinet.com/mantrip, July 2002.
4 Information society technologies.
5 Virtual private network.

decreases PDP complexity by automatically adapting network component behavior with environment changes.

4.3.7. *Dynamic service provisioning for mobile users*

Ganna and Horlait proposed an architecture based on mobile agents and the policies to manage users, their mobility and to provide them with required QoS and security [GAN 03].

4.3.7.1. *Main system components*

Manager agent: its responsibility is to create, destroy and add agents. It also uses policies to manage agent behavior.

Coordinator agent: it coordinates the interaction and information exchanges between agents.

Resource agent: it collects information on QoS resources and technologies from network elements and stores it in the *devices capabilities repository*. This information will be used by the manager agent for negotiating a security service or a QoS and for translating high level policies into network policies.

Authentication agent: its role is to control and identify users. If the user does not belong to its domain, it will go to his main domain to retrieve his profile. Otherwise, it will consult the *user's profiles and services repository*. This information database contains the profiles of users with a contract with their domain as well as the services provided by the domain with their characteristics and costs.

Policy agent: it assists the administrator in adding, editing and deleting policies. It also controls domain policy consistency stored in the *policy repository*. It provides the policies requested by the manager agent.

Security agent: it collects security information and network component capacities. It then sends the information back to the manager agent which stores it in the *devices capabilities repository*. This information is then used by the negotiation agent.

Services agent: it provides services to the user. If the requested service goes through another domain, it moves to the other domain to ensure service continuity.

Negotiation agent: if a domain cannot provide the service requested by a user, if the service must use another domain, or if the user does not belong to the domain, a negotiator agent is created to negotiate the service. It will move to the domains involved and once an agreement is reached, it informs the manager agent.

Monitor agent: it verifies that the negotiated contract with the user is respected. It sends a report to the managing agent if there is degradation or modification of parameter values.

Invoicing agent: it is responsible for invoicing the domain's clients for the service.

Policies play an important role. They manage the behavior of the entire system, they control agent life cycles, they indicate what has to be done when a user visits a domain and requests a service, they also guarantee that the established service will be maintained.

4.3.8. *Intelligent agents for dynamic security control*

The use of policies to manage IP network security is not a new concept [HEI 99, LUP 97, DAM 00]. However, this can become complex if users are mobile because the number of users changes constantly and user profiles are very diversified. Since agents are well adapted to dynamic and complex environments, using them will enable an easier management of security and especially to detect intrusions.

Boudaoud *et al.* propose an MAS-based model for network security management. Security policies guide MAS behavior [BOU 01]. They can specify what attack profiles to detect and which actions to take when an attack is detected.

Network security management is broken down into three plans:

– *user plan* where security policies are defined and specified by the administrator. Also at this level, security policies are modified following a network configuration change, for example, or new policies can be proposed to detect new attacks;

– *MAS plan* corresponding to the intelligent part of the system. The role of MAS is to identify security rule violations and to recognize attacks which may happen in the network. In order for the MAS to detect these attacks, a language for the representation of attack patterns was proposed; the agents analyze security events characterizing security attacks. To do this, they use a BDI (*beliefs*, *desires and intentions*) model;

– *network plan* made up of elements to secure: it is at this level that are created the security events which will be analyzed for the purpose of detecting attacks and ensure that security policies specified by the administrator are enforced. It is also at this level that the system can act on the network (by configuring firewalls, for example).

Security policies are vital. They occur at three levels:

– they select and create the security attack patterns to detect;

– they create goals for the agents who are responsible for detecting these security attack patterns;

– they select the events to filter in order to recognize the security attack patterns created.

4.4. Agents and service contract negotiation

Agents can assist the end-user and service providers in negotiating their service contracts. They are adapted for this task which requires a quick decision-making [CEL 00] mainly because of their autonomy, communication, cooperation and mobility capacities [VER 99]. Before we continue by giving a few examples, we will briefly explain what are service contracts.

4.4.1. *Service contract*

The SLA (*service level agreement*) represents a service contract between a client and a service provider. The contract elements will indicate how the service will be rendered and the measures to take for effectively control the service.

There are two SLA levels:

– the client SLA: this is the SLA binding an end client to his service provider;

– the interdomain SLA: it is an SLA between two service providers each offering the same level of service (two IP domains, for example). The provider asking for a service does so on his client's behalf.

End-to-end SLA will only happen if each SLA is respected.

The QoS elements within an SLA client have to do with QoS such as it is requested by the user, i.e. end-to-end QoS. It will then be necessary to propagate the information in the different domains as shown in Figure 4.5.

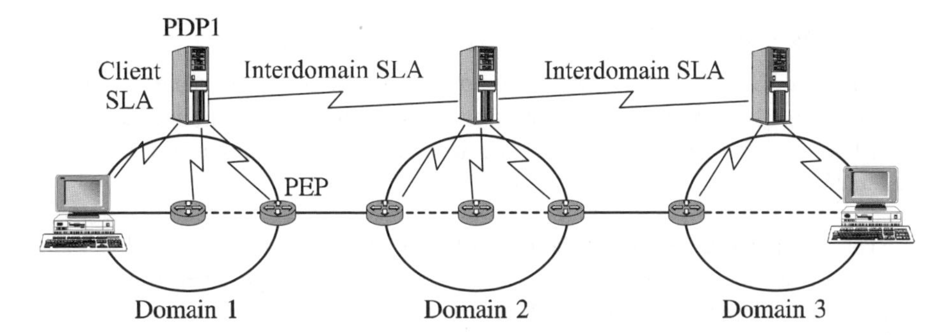

Figure 4.5. *SLA negotiation*

Each domain can implement different QoS mechanisms and belong to different providers. This increases the difficulty in propagating QoS and in finally ensuring the compliance with the client's SLA.

The SLS [TEQ 00] correspond to the technical part of SLA. It contains the technical description of services. It is important to standardize SLS contents in order to easily negotiate or renegotiate its terms and decrease the service deployment time. This standardization is not easy because the SLS can apply to a variety of services. Today the most advanced studies in this field are those by the EURESCOM [EUR 01] and IST-TEQUILA project [TEQ 00].

Later on, we will present a dynamic interface for the negotiation of SLS parameters whose functionalities have been extended to choosing the best service provider. We will also describe two architectures enabling the dynamic negotiation of intra and interdomain SLAs: one for fixed users and the other for mobile users.

4.4.2. *An intelligent negotiation interface*

Agents are often called intelligent interface or interface agent when their main goal is to collaborate with a user. They can assist users by executing tasks on their behalf asynchronously or by hiding the complexity of difficult tasks. The number of tasks that they can accomplish is unlimited.

We will present here the studies undertaken for project RNRT[6] ARCADE concerning the definition and achievement of an intelligent interface for the negotiation of QoS parameters.

4.4.2.1. *Project RNRT ARCADE*

The goal of the RNRT ARCADE project (IP environment adaptive control architecture) is to define a control and management architecture in third generation IP networks and a certain number of services including the automation of service negotiation.

The architecture retained is a policy-based control architecture with a new direction from the one proposed by IETF: control does not go from user to edge router and then from edge router to PDP. Instead, it goes directly from the terminal to PDP. There are many advantages to this architecture:

– each terminal partly shares its power with the network which results in a large network cost reduction;

– security can be handled from the terminal without waiting to reach the edge router;

– better consideration of user needs;

– mobility with QoS is much easier to take into account because it is managed by the terminal and not by the machine on which there will be a connection;

– intelligence is mainly located in user hardware and PDP, which reduces router complexity.

COPS-SLS protocol is used for direct negotiation between PDP and terminal. Since the process of negotiation is complex (the user or application must provide the technical parameters needed, renegotiate if necessary and manage the degradation by accepting or refusing the service provider's proposition), an intelligent interface was introduced in the station in order to (re)negotiate SLS parameters by taking into consideration user needs as well as the characteristics of the application launched. This intelligent interface, called NIA (*negotiation interface agent*), also measures QoS at the station and compares results with the level of service requested. This enables the user to have his own data as well as to renegotiate if needed.

The terminal is also equipped with a classifier which makes it possible to recognize and classify packets for each application. Once classified, the packets are

6 Réseau national de recherche en télécommunications (national network of research in telecommunications).

marked (DSCP fields). Marking depends on the negotiation result between the station and PDP SLS.

The NIA is an MAS which, by learning, must identify and associate SLS parameters for each application and user profile. To do this, it has a knowledge base, which is acquired throughout this understanding, which enables it to dynamically negotiate the service level with PDP SLS via PEP SLS.

4.4.2.1.1. Functional specification of intelligent interface

The MAS is made up of five agents. Each agent has a specific task which can be broken down into modules.

The user agent must identify the user profile. It is broken down into three modules: user identification, data storage and modification, and analysis of user behavior.

The application agent must determine the profile of the application launched by a user. The application profile is determined from the application class as well as the degree of importance of this application specified by the user. The applications are classified based on their QoS requirements (delay, throughput, etc.). The application agent is made up of two modules: classification of applications and determination of the application profile.

The parameters agent is responsible for determining a general negotiation profile based on the application and user profiles. This agent is made up of three modules: determination of SLS parameters, determination of predefined values and value filling.

The interface agent makes the connection between the user and the system, and it communicates with the user for information. If the user is known, the agent may only ask for the validation (or not) of its previous choices. The interface agent is made up of two modules: translation of requests and message formulation.

The control agent makes the connection between a terminal and the service provider. It is made up of three modules: communication, QoS evaluation and negotiation.

Once the service class is obtained, NIA determines the application filters and sends them to the classifier which will use them to mark the packets with the DSCP value obtained.

4.4.2.1.2. Development of an intelligent interface

The intelligent interface was developed from the oRis[7] multi-agent platform which was designed in the LIL-ENIB laboratory. It has also been tested during the full scale control architecture experiment of project ARCADE presented in Figure 4.6. In this context, the intelligent interface dynamically negotiates SLS audio and video application parameters on behalf of the user.

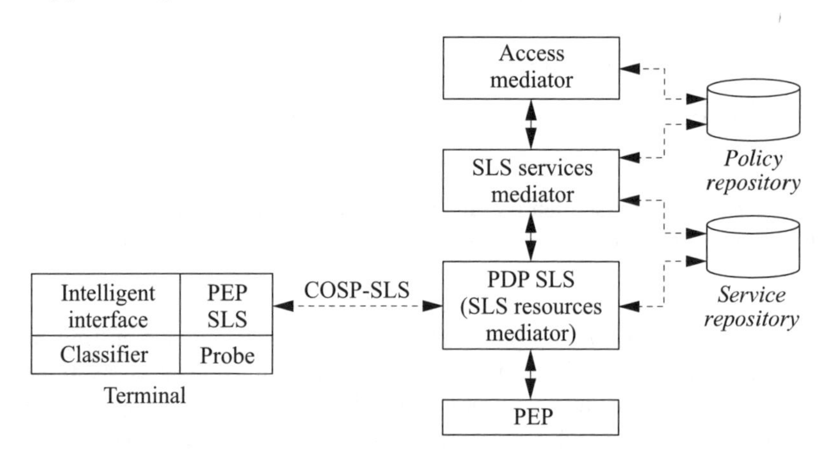

Figure 4.6. *ARCADE control architecture*

4.4.3. *Client-provider dynamic negotiation*

A new functionality was added to the intelligent interface that we have just described in order for it to be able to select the best service provider, always on behalf of the user [KLI 03]. For this, two new types of agents were introduced:

– *mobile agents* called UN (*user negotiators*) are sent towards access mediators which assume the role of *brokers* in order to negotiate service offers according to user requirements;

– *provider choice agent* called UO (*user overseer*) selects the best offer among service offers coming from mobile agents. These agents also send new offers which may interest the user.

The access mediator negotiates the best service offer on behalf of the provider with mobile agents via its AN (*access negotiator*) agent (Figure 4.7).

7 http://enib.fr.

Once the service provider has been retained, an SLA is established between the client and the service provider, via the access mediator retained.

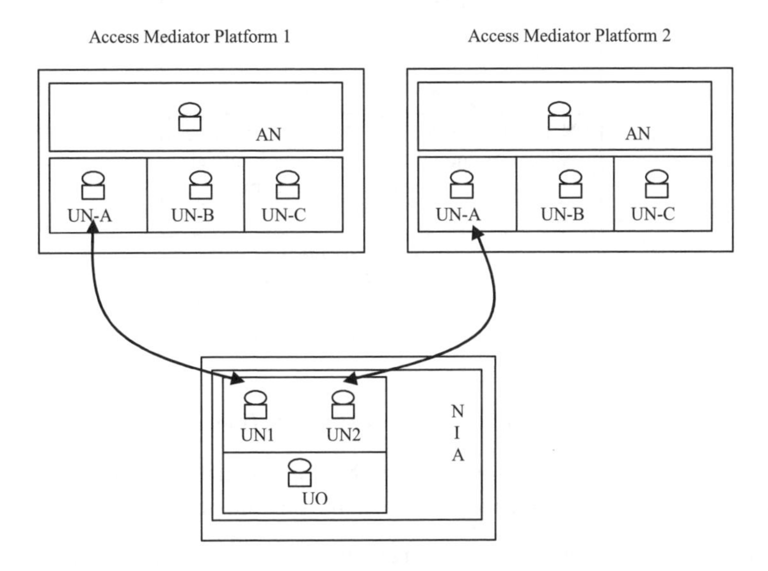

Figure 4.7. *Architecture agents*

Agents simplify a user's workload by dynamically negotiating SLS parameters in his place. They also offer the user more freedom in changing service providers. In addition, the use of mobile agents limits communication costs and enables asynchronous communications while ensuring a certain degree of confidentiality.

4.4.4. *Dynamic negotiation between providers*

In [FON 03], the authors proposed to pair the policy-based management approach with mobile agent technology to establish SLAs with surrounding ISPs and to control communications.

The architecture proposed enables the negotiation process between a client and his ISP and between ISPs. It is based on a set of fixed and mobile agents. Fixed agents are used within a domain to interact with the local policy-based management system for the negotiation of an SLA. Mobile agents move from one policy-based management system to another to negotiate an SLA on behalf of the initiator (client or ISP). They transport high level policies in such a way that the negotiation and decision can be handled locally.

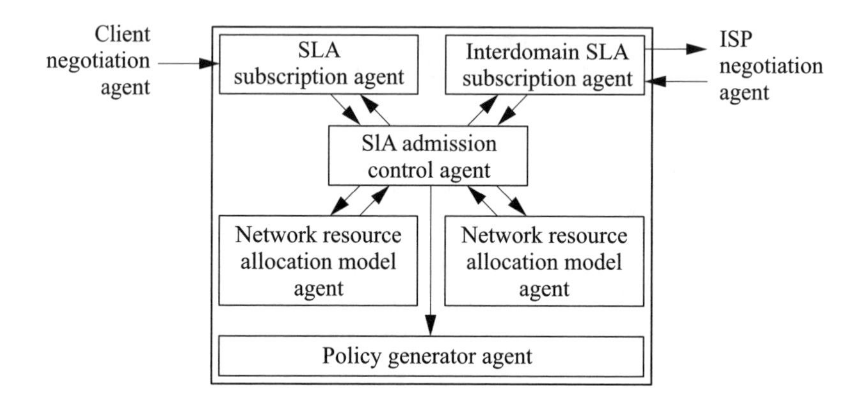

Figure 4.8. *ISP architecture*

The system contains the following types of agents (Figure 4.8):

– the *client negotiation agent* is responsible for negotiating with the client's ISP. It transports the SLA request as well as client negotiation policies;

– the *SLA subscription agent* is responsible for interacting with the client negotiation agent for the subscription of a new contract. It sends the request to the SLA admission control agent after some initial verifications such as client identity;

– the *interdomain SLA subscription agent* is responsible for SLA negotiation with surrounding ISPs. To do this, it uses an ISP negotiation agent. In return, it sends the SLA to the SLA admission control agent;

– the *SLA admission control agent* verifies if the new contract can or cannot be accepted according to the ISP's policy and possibilities, available resources and existing contracts. It interacts with the interdomain SLA subscription agent to verify that the new contract terms are compatible with the resources available in the domain and between domains. It obtains available resources from the different domains involved with the help of network resource allocation model agents (intra and interdomain) and in return it sends to them and to the policy generator agent the terms of the accepted contract;

– the *network resource allocation model agent* updates information concerning the resources available in the ISP network;

– the *interdomain network resources allocation model agent* updates information concerning the other ISPs;

– the *policy generator agent* is mainly responsible for translating accepted contracts into operational policies and for storing them in the *policy repository*;

– the *ISP negotiation agent* is created by the interdomain SLA subscription agent to negotiate a contract with its neighboring ISP. The negotiation takes place if there is no existing inter-ISP resource allocation or if the resources available for inter-ISP communications are not enough to satisfy the demand.

The automation of negotiation and configuration procedures enables a quick introduction of end-to-end service by relying on negotiated SLAs. The policy-based management approach facilitates ISP strategies representation through policy rules. The complexity of the negotiation process is left to mobile agents, which also leads to a reduction in the number of exchanges. It is obviously a significant approach in the automation of the negotiation process.

The agent approach makes it possible to break down the management system into components. They then can easily be combined with the underlying policy-based management components.

4.4.5. *Dynamic services negotiation for mobile users*

The architecture proposed by Ganna and Horlait and presented in section 4.3.7 defines the entities required to manage mobile users and provide them with end-to-end services with guaranteed QoS and security. It is based on the notion of domain. If a domain cannot provide the service requested by a user, if the service involves another domain, or if the user does not belong to the domain, *a negotiating agent* is created to negotiate the service. It then moves to the domains involved. Negotiation includes the mobile user and the different domains. The information contained in the databases (*user's profiles and services repository*, *devices capabilities repository*, *policy repository*) guide the negotiating agent in its negotiation. In this way, the user profile serves as a starting point for the negotiation. Once an agreement is reached, it informs the managing agent.

4.5. Management of emerging services

With ambient Internet, users will enter a totally nomadic era because they will be able to connect on the Internet from anywhere in the world. This will cause a change in their life and work behavior since they should be able to work as if they were in their office. An electronic market of services ranging from simple communication services to complex distributed multimedia applications should quickly start becoming available [MAG 96].

An agent-based architecture and policy-based management will enable the creation, procurement and management of personalized services as well as transparent access to resources for the nomadic users. In order for personalization to be effective, we must have access to a rich set of context data, i.e. to information relative the profile, presence, location, identity of users and services, available services and resources, local transmission capabilities, workstation capacities, time, etc. From this context information and other information, in particular relative to company policy, the system must be able to release a certain level of knowledge in order to:

– propose or compose personalized services;

– establish and maintain virtual workgroups based on common interests. These groups will have the possibility to share or reserve resources that will be attributed to them according to a profile.

We present two architectures making it possible to propose and manage new services in a flexible way. First, we will explain what we mean by emerging service.

4.5.1. *Emerging services*

Emerging services have the capacity to adapt to a context. Dey *et al.* give the following definition of context [DEY 00]:

> Context is any information that can be used to characterize the situation of an entity. An entity can be a person, place, or physical or computational object which occurs in the interaction between a user and an application. It includes the user and the applications.

Yan *et al.* reflect on the next generation network (*network-centric*) and propose the following definition [YAN 03]:

> Context is any information, which is obtained either explicitly or implicitly, that can be used to characterize a certain aspect of an entity that is involved in a specific application or network service. An entity can be a physical object such as a person, a place, a router, a 3G network gateway, a physical link, or a virtual object such as an IPsec tunnel, or even an SNMP agent.

Context information must be represented, stored and maintained. It is complex because it is changing. A personal agent understanding the needs of a user could interact with other agents to constitute context knowledge [KAR 03]. The use of context information also requires a complex decision process which can be facilitated by the introduction of intelligent agents.

We will see later on that combined with policy-based management, intelligent agents offer the possibility of forming architectures enabling the management of emerging services in a quick, flexible and automatic way.

4.5.2. *Dynamic management of emerging services*

We present here studies that were part of the IST CONTEXT project[8].

IST CONTEXT project

Yan *et al.* have proposed an architecture for managing emerging services [YAN 03] as part of this project. Location, identity, time and activity are considered as basic context types to characterize the situation of a specific entity. Policies serve as context information containers. A context information model based on PCIM and its PCIMe extensions was proposed as well as a scenario named "*super-mother*".

Super-mother scenario

In this scenario, a user (a mother working from home and having to go to the hospital because her child is sick) with a subscription to a *context-aware* service will benefit from a service adapted to her needs (this mother must urgently send a large file to work). In this example, the access point of the hospital's WLAN will signal the arrival of a user with a *context-aware* service subscription. A mobile agent will send policies to the element's managing station in order to configure the access point and enable the user to access the needed service. Since the communication must be secure, the managing station will send mobile agents that will configure a secure IP tunnel between both routers (in this case the machine on which the access point is installed is the company's access router). This tested scenario has demonstrated the feasibility of the method in the next network generation.

The provisioning of emerging services is made easier by mobile agents who are controlled by policies. The introduction of agents offers more flexibility and autonomy to the emerging service in order to respond to changes in environment such as location, workstation size and network characteristics without user disruption.

---

8 CONTEXT project website: http://www.context.upc.es (IST-2001-38142-CONTEXT).

4.5.3. *Dynamic management of group multimedia services*

[HAR 01] has developed a multimedia environment for the constitution of virtual teams.

A virtual team groups a dynamic collection of individuals, a set of common services and network facilities. The virtual team shares resources in a coordinated, flexible and secure way.

The system architecture is made up of several agents cooperating to provide an environment to virtual teams. There again, an agent's behavior is guided by a set of policies.

The key component of this system is the *context agent*. It manages information from virtual team participants such as their capacities, roles, services used by all or some of its members, resources required during team creation and network services. From this information, the agent generates a set of policies monitoring the behavior of agents representing virtual team participants. The context agent is responsible for the automatic authentication of participants via their personal assistant. Personal assistants are the agents representing participants and act on their behalf. The context agent provides an interface which makes it possible for an authorized authority (a team manager, for example) to create a virtual team by specifying its members with their privileges, the resources assigned to the team as well as the QoS assigned to different team services. To start a session, a negotiation process through personal assistants can be started and controlled by a context agent who plays the role of mediator. The context agent is made up of three components:

– a policy generator which implements and interprets context information;

– a team manager which provides access to context information;

– a team assistant which uses the context policies to launch events that apply to specific agents based on their state and usage context.

When an authorized authority (a team leader, for example) creates a virtual team, the system assigns a *context agent* to manage this particular environment and a *network service agent* to support its network services. A *control agent* will then be associated with each session and activated by the team's participants. These three agents establish a relation between participants, services or applications and network resources.

Agents enable the quick and robust development of reusable applications. The use of policies greatly reduces system complexity and enables an efficient control of the virtual team's activities.

4.6. Conclusion

Policies are viewed as a way to guide network behavior through a high level declarative language. This keeps the administrator from having to manually configure each component. Policy-based management also offers more flexibility because the administrator can configure network elements by simply entering or changing policies.

Agent technology provides a way to manage networks in a more flexible and autonomous way. Agent intelligence levels depend on tasks to execute or on their attributed roles.

Integrating agent technology and policy-based management would make it possible to automatically adapt the behavior of network components to environment changes. Networks would operate in a consistent way without human intervention. Both of these approaches thus provide a response to the demand of the new highly dynamic generations of networks.

Policy-based management and agent technology go well together. On the one hand, agents can build policy-based management in a flexible way. On the other hand, policies are used to guide agent behavior. In this context, Hamada *et al.* speak of active policies [HAM 00].

In order to encourage the expansion of agent utilization, we must resolve the security problem because their autonomy, mobility and intelligence properties make the use of agents complicated. Policies could be used to ensure that agent behavior remains coherent.

It is also important to ensure that the new policies do not compete with existing policies. A number of studies have been conducted on this theme, some going as far as calling for reactive agents [WIT 02].

An interesting research subject could be a proposition of distributed architectures combining intelligent agents and policy-based management for autonomous management [KEP 03].

4.7. Bibliography

[AFM 00] AGOULMINE N., FONSECA M., MARSHALL A., "Multi domain policy based management using mobile agent", *Lecture Notes in Computer Science*, 2000.

[BAL 95] BALANDIER R., CAMINADA A., LEMOINE V., ALEXANDRE F., "170 MHz field strength prediction in urban environment using neural nets", *IEEE PIMRC'95*, vol. 1, p. 120-124, September 1995.

[BAL 98] BALDI M., PICCO G., "Evaluating the Tradeoffs of Mobile Code Design Paradigms", *Network Management Applications*, 1998.

[BAR 01] BARRÈRE F., BENZEKRI A., GRASSET F., LABOREDE R., RAYNAUD Y., "Distribution de politique de sécurité Ipsec", *GRES'01*, Marrakech, Morocco, December 2001.

[BHM 98] BREUGST M., HAGEN L., MAGDANZ T., "Impacts of mobile agent technology on mobile communications system evolution", *IEEE Personal Communication Magazine*, August 1998.

[BLA 98] BLAKE S., BLACK D., CARLSON M., DAVIES E., WANG Z., WEISS W., "An Architecture for Differentiated Services", *RFC 2475*, December 1998.

[BRI 01] BRIOT J.P., DEMAZEAU Y., *Principe et architecture des systèmes multi-agents*, Hermes, 2001.

[BOU 01] BOUDAOUD K., GUESSOUM Z., MCCATHIENEVILE C., DUBOIS P., *Policy-based Security Management using a Multi-agent System*, 2001.

[BOU 02] BOUKHATEM N., CAMPEDEL B., CHAOUCHI H., GUYOT V., KRIEF F., NGUYEN T.M.T., PUJOLLE G., "I3 – A New Intelligent Generation for Internet Networks", *Smartnet'2002*, IFIP WG 6.7, Kluwer, Saariselkä, Finland, April 2002.

[BOU 95] BOUTABA R., XIAO J., "Network Management: State of the Art", *IEEE PIMRC'95*, vol. 1, p. 120-124, September 1995.

[BOY 00] BOYLE J., COHEN R., DURHAM D., HERZOG S., RAJA R., SASTRY A., "The COPS (Common Open Policy Service) Protocol", *RFC 2748*, 2000.

[CAR 00] CARDON A.E., *Service level specification for interdomain QoS negotiation*, November 2000.

[CEL 00] CELENTI E., RAJAN R., DUTTA S., *Conscience artificielle et systèmes adaptatifs*, Eyrolles, 2000.

[CHA 01] CHAN K., SELIGSON J., DURHAM D., GAI S., MCCLOGHRIE K., HERZOG S., REICHMEYER F., YAVATKAR R., SMITH A., "COPS Usage for Policy Provisioning (COPS-PR)", *RFC 3084*, 2001.

[DAM 00] DAMIANOU N., DULAY N., LUPU E., SLOMA M. "A Language for Specifying Security and Management Policies for Distributed Systems", *Imperial College Research Report*, Doc 2000/1, Department of Computing, Imperial College of Science Technology and Medicine, London University, 2000.

[DEY 00] DEY A.K., ABOWD G.D., "Towards a better understanding of context and context-awareness", *Workshop on the What, Who, Where, When and How of Context-Awareness*, April 2000.

[EUR 01] EURESCOM P1008, *Interoperator interfaces for ensuring end to end QoS. Selected scenarios and requirements for end-to-end IP QoS management*, January 2001.

[FER 99] FERIDUN M., KASTELEIJN W., KRAUSE J., "Distributed management with mobile components", *IEEE IM'99*, October 1999.

[FES 01] FESTOR O., "Ingénierie de la gestion de réseaux et de services: du modèle OSI à la technologie active", Dissertation on authority to direct research, Nancy University 1, 2001.

[FIP 01] FIPA, "FIPA Policies and Domains Specification", *FIPA TC Architecture*, PC00089D, August 2001.

[FLE 02] FLEGKAS P., TRIMINTZIOS P., PAVLOU G., "A Policy-Based Quality of Service Management System for IP DiffServ Networks", *IEEE Network, Special Issue on Policy-Based Networking*, p. 55-56, March-April 2002.

[FON 03] FONSECA M., AGOULMINE A., *Policy Based Mobile Agents as a Solution for Intercarriers SLA Negotiation to Support End-to-End IP QoS*, 2003.

[GAN 03] GANNA M., HORLAIT E., "Policy-Based Service Provisioning and Users Management Using Mobile Agents", *MATA'2003*, Marrakech, Morocco, October 2003.

[GAY 02] GAY V., DUFLOS S., KERVELLA B., DIAZ G., HORLAIT E., "Policy-Based quality of service and Security Management for Multimedia Services on IP networks in the RTIPA project", *MMNS 2002*, Santa Barbara, USA, 2002.

[HAM 00] HAMADA T., CZEZOWSKI P., CHUJO T., *Policy-based Management for Enterprise and Carrier IP Networking*, 2000.

[HAR 01] HARROUD H., LAKHDISSI M., KARMOUCH A., GROSSNER C., "Policy-Based Management for Multimedia Collaborative Services", *MMNS'01*, 2001.

[HEI 99] HEILBRONNER, "Requirements for Policy-Based Management of Nomadic Computing Infrastructures", *HP-OVUA'99*, Bologna, Italy, June 1999.

[JRA 03] JRAD Z., KRIEF F., BENMAMMAR B., "An Intelligent User Interface for the Dynamic Negotiation of QoS", *ICT'2003*, Papeete, Tahiti, February 2003.

[KEP 03] KEPHART J., CHESS D., "The Vision of Automatic Computing", *IEEE Computer Society*, January 2003.

[KLE 03] KLEIN G., KRIEF F., "Mobile agents for Dynamic SLA Negotiation. International Workshop on Mobile Agents for Telecommunication Applications", *MATA'2003,* Lecture Notes on Computer Science, Marrakech, Morocco, October 2003.

[KRA 96] KRAUSE S., MAGEDANZ T., "Mobile service agents enabling intelligence on demand in telecommunications", *GLOBECOM'96*, 1996.

[LYM 02] LYMBEROPOULOS L., LUPU E., SLOMAN M., "An Adaptive Policy Based Management Framework for Differentiated Services Networks", *POLICY'02*, Monterey, California, p. 147-158, June 2002.

[LUP 97] LUPU E., SLOMAN M., "Conflict Analysis for Management Policies. IFIP/IEEE International Symposium on Integrated Network Management", *IM'97*, San Diego, USA, May 1997.

[MAR 66] MAR D.A., SLOMAN M. "Implementation of a Management Agent for Interpreting Obligation Policy", *IEEE/IFIP DSOM'96*, Aquila, Italy, October 1996.

[MOO 02] MOORE B., DURHAM D., STRASSNER J., WESTERINEN A., WEISS W., HALPERN J., "Information Model for Describing Network Device QoS Datapath Mechanisms", *Internet draft*, May 2002.

[MOO 01] MOORE B., ELLESSON E., STRASSNER J., WESTERINEN A., "Policy Core Information Model", *RFC 3060*, 2001.

[PUJ 04] PUJOLLE G., *Les réseaux*, Eyrolles, 2004.

[SAM 03] SAMAAN N., KARMOUCH A., "A multi-Agent Framework for QoS adaptation Using Policies in *wireless* Environments", *GRES'2003*, Lecture Notes on Computer Science, Marrakech, Morocco, October 2003.

[SCH 94] SCHILIT B., THEIMER M., "Disseminating Active Map Information to Mobile Hosts", *IEEE Networks*, 8(5), p. 22-32, 1994.

[SER 03] SERGOIU C., ARYS G., WOOLDRIDGE M., "A Policy Based Framework for Agents: On the Specification of an Agent Policy Language including Roles, Relationships, Conversation Patterns and Cooperation Patterns", *AAMAS'2003*, p. 1126-1127, 2003.

[SNI 01] SNIR Y., RAMBERG Y., STRASSNER J., COHEN R., MOORE B., "Policy QoS Information Model", *Internet Draft*, November 2001.

[TEQ 00] TEQUILA, "SLS: Service Level Specification Semantics and Parameters", *Internet Draft*, November 2000.

[VER 99] VERMA D., *Supporting service level agreements on IP networks*, Macmillan Technical Publishing, 1999.

[WAG 02] WAGNER T., "An agent-oriented approach to industrial automation systems", *MALCEB'02*, 2002.

[WIT 02] WITTNER O., HELVIK BJARNE E., "Robust Implementation of Policies Using Ant-like Agents", *Net-Con'02*, Paris, October 2002.

[WOO 95] WOOLDRIDGE M., JENNINGS N.R., "Intelligent Agents: Theory and Practice", *The Knowledge Engineering Review*, vol. 10, no. 2, 1995.

[YAN 02] YANG K., GALIS A., MOTA T., GOUVERIS S., "Automated Management of IP Networks Through Policy and Mobile Agents", *MATA02*, p. 249-258, Barcelona, 2002.

[YAN 03] YANG K., GALIS A., TODD C., "Policy-driven Mobile Agents for Context-aware Service in Next Generation Networks", *MATA'2003*, Lecture Notes on Computer Science, Marrakech, Morocco, October 2003.

[YAV 00] YAVATKAR R., PENDARAKIS D., GUERIN R., "A Framework for Policy Based Admission Control", *RFC 2753*, 2000.

Chapter 5

Multi-agent Platforms

5.1. Introduction

A large number of applications require the development of complex systems distributed on several interacting computers which must communicate between each other while executing independently, sometimes even as sub-systems or sub-tasks leading to a common goal. Traditional methodologies such as object-oriented do not provide adequate concepts to master the complexity of distributed systems, in particular modeling systems in which connections dynamically evolve. In order to compensate for these inadequacies, object-oriented programming has recently evolved into a programming which enables the natural design of this type of system. Several methodologies constituting an extension of object-oriented or knowledge-based methodologies were proposed, in particular for the development of multi-agent systems (MAS). These are systems made up of a set of autonomous entities called agents. Currently, an increasing number of applications are adopting MAS. This requires the introduction of standards controlling the agent-oriented programming and methodologies, as well as the development of adequate modeling techniques.

In addition, MAS are information technology fields where software is the most difficult to develop because of the large number of concerns to consider: communications, resource management, other agent searches, etc. The concept of platform, which is linked to the implementation of MAS, is an environment in which agents can evolve. Platforms are environments for managing agent life cycles and the services to which they have access. Moreover, few efforts were made in the

---

Chapter written by Zeina EL FERKH JRAD.

standardization of MAS platforms. It seems obvious that MAS development still remains an open field.

The goal of this chapter is not to compare the various environments of MAS development, nor to make a better classification, but rather to present specific MAS platforms which we consider to be interesting in a network context.

5.2. Towards a standardization of multi-agent technology

For several years now, researchers as well as manufacturers have carried out several studies on agents. However, few attempts were focused on the harmonization of SMA architectures. This lack of consensus is partly caused by the absence of a common design in research circles of the main general principles on which SMA architectures must be based [RIC 00]. It is only recently that several groups of independent manufacturers and researchers tried to propose a standardization of multi-agent technology. Some of these groups are:

– Foundation for Intelligent Physical Agents (FIPA);

– Knowledgeable Agent-oriented System (KAOS);

– General Magic Group.

5.2.1. *FIPA model*

FIPA[1] is a multidisciplinary group focusing on standardizing agent technology. Its mission is to facilitate the interoperability between agents and MAS from different providers [SAB 01]. Since 1997, FIPA has developed a set of specifications ranging from communications languages (*agent communication languages*) to content languages and to interaction protocols (Figure 5.1). The FIPA-ACL language follows the KQML style (using performance factors resulting from the theory of language acts and other parameters), albeit with better identified semantics. The language also considers the use of interaction protocols.

1 http://www.fipa.org.

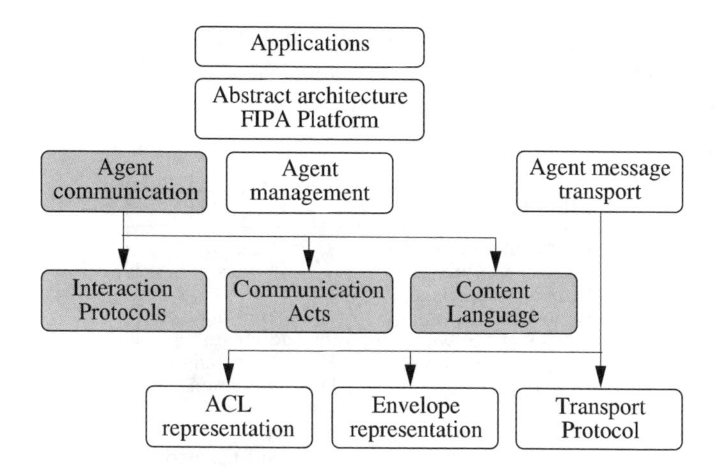

Figure 5.1. *FIPA specifications*

FIPA specifies the interfaces of the various environment components with which an agent can interact (i.e. human, other agents, other software) and the physical world. FIPA has developed two types of specifications [BOI 01]:

– "formative" to describe the agent's external behavior and to ensure interoperability with other specified FIPA subsystems;

– "informative", for applications representing a guide for the industry concerning the use of FIPA technology.

To manage agents in an open environment, FIPA's approach for MAS development is based on a paradigm described by using a reference model defining a normative environment in which agents exist and operate, and an agent platform indicating an infrastructure for the deployment and interaction of agents.

The platform proposed by FIPA contains three main components which are evaluated during interoperability tests: the message transport system, agent management system and directory facilitator. The protocol used is Iiop and the message structure is defined with accuracy.

5.2.1.1. *Illustration*

Figure 5.2 illustrates the structure of a FIPA message as well as the transmission of a message between two agents located on separate platforms. The exchange of messages occurs between two A and B agents. Agent A builds the body of the message, which represents the exchange semantics and the envelope that gathers transport information (encoding, protocol, etc.). It then delegates the transmission of the message to the *message transport system* which, according to the protocol used (Iiop, HTTP, Rmi, etc.), will select a *message transport provider* and communicate with the platform hosting agent B.

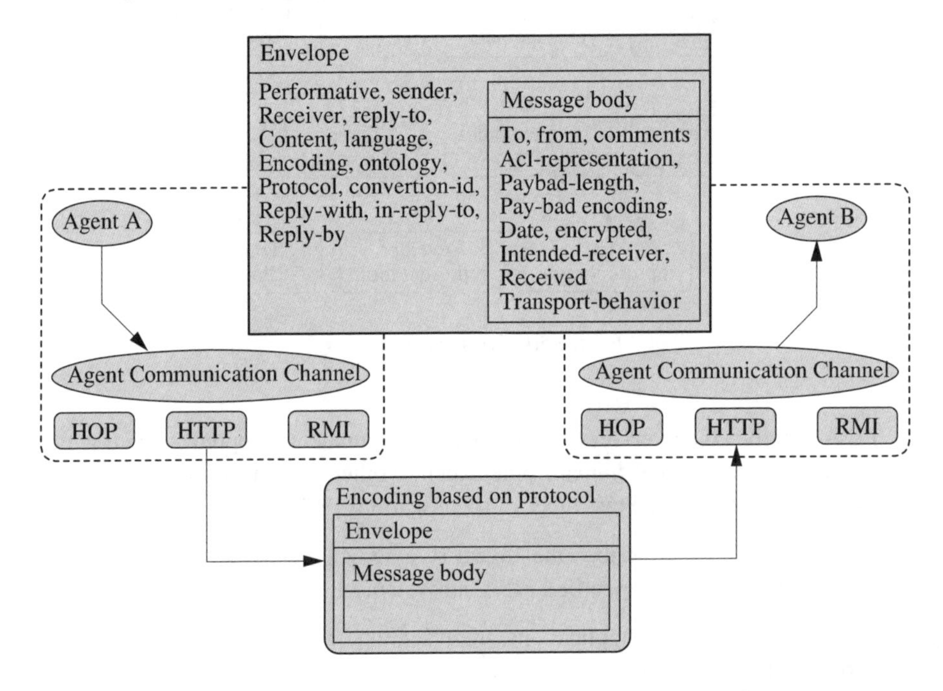

Figure 5.2. *Message exchange between FIPA agents A and B*

The concept of platform is used as an agent management and communication infrastructure. These functionalities are transported through accessible service platforms either to platform agents or to other platforms.

Table 5.1 shows some examples of platforms which are compliant with FIPA specifications.

Product	Research institute	Reference
Agent Development Kit	Tryllian BV	http://www.tryllian.com
Zeus	British Telecom	http://193.113.209.147/projects/ agents/zeus
FIPA-OS	Emorphia	http://fipa-os.sourceforge.net
April Agent Platform	Fujitsu	http://sf.us.agentcities.net/aap
Grasshopper	IV++ Technologies AV	http://www.grasshopper.de
JACK Intelligent Agents	Agent-Oriented Software	http://www.agent-software.com
Java Agent Development Environment (Jade)	TILAB (CSELT)	http://sharon.cselt.it/projects/jade
Lightweight Extensible Agent Platform (Leap)	Motorola	http://leap.crm-paris.com

Table 5.1. *Examples of platforms which are compliant with FIPA specifications*

5.2.2. *KAOS model*

The KAOS (*knowledgeable agent-oriented system*) group has proposed an open and distributed MAS architecture for software agents. The main goal of KAOS [BRA 96] is to deal with the two major limitations of agent technology. The first problem is handled by using commercial distributed object offerings (CORBA, DCOM, Java), as a foundation for agent functionality and by supporting standardization research and attempts to resolve agent interoperability problems. The second problem is dealt with by developing a communication agent meta-architecture in which any language of communication combined with its semantics can be adapted. Contrary to the majority of communication agent architectures, KAOS explicitly takes the individual message into account as well as the various message sequences in which it can happen. The architecture describes agent implementations (from the notion of simple agent to the notion of agent role, as with mediators, for example) and develops communication messages by using conversation policies on the basis of agent-to-agent dynamic interaction.

5.2.3. *General Magic model*

General Magic is an attempt at commercial research on mobile agent technology for the electronic trade. General Magic Telescript language [WHI 96] is a rich object-oriented language which hides many aspects of mobility, security, object duration and process interactions from programmers. It is based on mobile agents moving from one server to another (telescript "places") and can interact locally with applications hosted by the server. This infrastructure makes it possible to develop electronic commerce applications where an agent can carry out sophisticated requests on behalf of a user by going from one place to another and interacting with the applications (mediators, service providers, etc.) that they host, following the diagram contained in its telescript code.

Conceptually, this technology models an MAS-like electronic trade where suppliers and consumers can meet and do business. Mobile agents are represented as entities located in a specific place at a given moment and this market is modeled as a network of computers supporting a set of locations which offer services to mobile agents. Below is a description of agent skills:

– they can move from one location to another and they have the right to travel;

– they can meet and launch procedures from other agents;

– they can create connections to communicate with an agent from another location;

– they are representative of their individual physical world or the organization that they represent;

– they have licenses indicating their skills.

5.3. Characteristics of a multi-agent platform

Being autonomous, in interaction and located in an environment are among the known fundamental characteristics of agent capabilities. However, these characteristics do not define all the possible types of agents such as, for example, BDI agents which imply the addition of a certain number of *cognitive* characteristics to agents, or agents communicating with languages such as KQML which require additional functionalities.

It is therefore not the platform's role to propose all possible types of agents, but rather only a set of functionalities. Figure 5.3 shows how the developer can combine these functionalities at will to generate several types of possible agents.

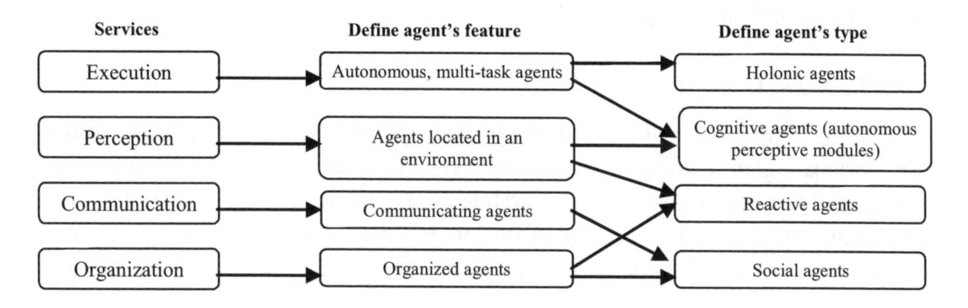

Figure 5.3. *Types of agents developed from services*

In order to achieve a large set of agents, Pierre [PIE 03] makes the assumption that it is enough to define a set of basic functionalities and a mechanism of extension. In this way, it defines the requirements necessary to build a platform enabling MAS.

5.3.1. *Methodological requirements for a multi-agent simulation platform*

With the objective of taking part in the development of a general structure which characterizes an ideal platform type[2], [SIM 01] targets three different objectives. The first objective is to define a reference model for the specification of multi-agent simulation platform requirements. The second is to establish a comparative analysis between various available platforms for the research committee. The third objective is the specification, design and implementation of a platform for accomplishing part of these requirements focusing on agent simulation. Nuno *et al.* [NUN 02] report their results for the second and third objectives. Marietto *et al.* [MAR 02] focus on the second objective and highlight requirements for an agent simulation platform. These requirements revolve around the following axes.

5.3.1.1. *Technical axis*

According to this axis, a simulation platform must provide services enabling the use of operating systems and their associated resources, in particular the network. It must also offer controlled execution and reproducible simulation services.

2 http://www.lti.pcs.usp.br/SimCog.

5.3.1.2. *Domain axis*

The platform must be able to propose a set of agent models to guide the design of agents. It must also offer fault management (for example, corrupted messages) at the technical and behavioral simulation level.

5.3.1.3. *Development axis*

The platform must make it possible to manage communication methods and to use various agent communication languages. It must also make reactive and cognitive agents available to agent architecture models. Lastly, one of the most important aspects is that the platform must offer abstractions for the agent organization and must be able to manage several sets of agents.

The two remaining axes developed in [MAR 02] are analysis and exploration. They imply that agents are cognitive and therefore too specific for the needs of agent-based simulations used in the scientific field. In spite of these differences, the three axes (technical, domain and development) seem completely relevant for the design of a general platform. We should mention that a platform can influence the agent-based system design; for example, the MadKit platform [IRON 89] offers its organizational metamodel Aalaadin, but it is not a requirement. In order to give more freedom to the developer, it would be even better if the platform did not impose a methodology or a too restrictive model. However, the developer should never be abandoned: the platform must also guide the developer in the system's analysis and design.

5.3.2. *Other forms of requirements for an agent platform*

The requirements suggested above are based on methodological aspects (technical, domain, development, analysis, etc.). The disadvantage of this methodological direction is that it is difficult to represent the various entities which are necessary to the realization of an agent system. Pierre [PIE 03] proposes a classification of agent platform requirements based on various entities:

– *agents* being the components of the system to create;

– the environment representing resources of the system to be created, which also includes agents;

– the platform which is the software making it possible for agents to evolve in the environment autonomously and to interact with each other.

It is necessary at this point to give a more precise definition of the roles of each element of the agent platform architecture proposition described above in order to complete the classification of previously stated requirements. A requirements list for an agent platform can be defined as follows:

– agents:

  - accomplishment of the autonomy characteristic,
  - accomplishment of the interaction characteristic,
  - provide a set of predefined types of agents,
  - enable the development of specific types of agents;

– environment:

  - make it possible for agents to organize themselves,
  - enable agents to reach system resources,
  - provide transparent access to resources whatever their physical location;

– platform:

  - provide an infrastructure where agents can survive,
  - provide an infrastructure for managing the environment,
  - enable the specialization of existing functionalities or add new ones.

We could also broaden these requirements with notions of security and migration. If we want to be able to create mobile agents, a security mechanism must be made available by the platform in order to restrict access to the environment resources. This security service would rely on a protection mechanism managed by the environment. The migration of agents would also be provided by a service enabling agent persistence (because migration implies the use of persistence) which must also be taken into account in the platform.

5.4. Multi-agent platform evaluation

The development of criteria for evaluating multi-agent platforms was the subject of several research studies. Garneau [GAR 02], for example, has chosen a range of 0 to 4 and has given a rating which reflects the platform level based on the following evaluation criteria:

– tool-related methodology covers the various stages;

– ease of training for this tool;

– transition facility between development and implementation;

– tool flexibility;

– inter-agent communication;

– debugging tools;

– development support;

– MAS management support;

– reduction in the work involved and ease of implementation;

– database support;

– automatic code generation;

– scalability of code;

– deployment;

– documentation.

Another evaluation study from Ricordel *et al.* [RIC 00] discusses MAS considered as complex software requiring suitable methods such as the four stages of construction: analysis, design, development and deployment. These four steps are defined in the following way:

– analysis: the process of determining, separating and describing the type of problem and its domain. In practice it consists of identifying the problem's application and key;

– design: the process of defining the problem's solution architecture. It consists of specifying a main solution to the problem, for example, by using UML;

– development: the process of developing a functional solution to the problem. In practice, it consists of programming the solution by using a specific programming language;

– deployment: the process of applying the solution to a real problem.

He then determines four quality measures for each of the previous steps which are commonly involved with the practical development of MAS systems. These are:

– completeness: it specifies the quantity and quality of the documentation and tools provided;

– applicability: it defines the possibilities offered and the restrictions imposed;

– complexity: it includes the required developer's skill and the quantity of work necessary to complete the task;

– reuse: the work gained by reusing previous work.

These quality measurements applied to the previous development steps lead to several platform evaluation criteria such as:

– completeness of platform analysis;

– applicability of platform analysis;

– complexity of platform analysis;

– reuse of platform analysis;

– completeness of platform design, etc.

5.5. Examples of MAS platforms

We have presented above the research studies on multi-agent platform requirements either for standardization or as simple definition. However, all platforms do not necessarily use these criteria. The majority of them are designed based on either a specific agent model such as Zeus, or based on a particular application domain such as *Aglets* mobile agents, or *Swarm* simulation.

In fact, there are numerous multi-agent platforms dedicated to different agent models. We are not trying to give an extensive list of current platforms here. We briefly describe some platforms selected based on three different viewpoints:

– platforms for simulation such as Swarm, Cormas, StarLogo, Geamas, Mice, etc.;

– platforms for implementation such as Zeus, AgentBuilder, Jack, MadKit, DECAF, Magic, Mace, Mask, MoCAH, OSACA, etc.;

– platforms for mobility such as Aglets, Concordia, GOSIP, Grasshopper, Gypsy, JumpingBeams, GenA, KlAIM, Mole, etc.

5.5.1. *Platforms for simulation*

A simulation makes it possible to build an abstraction of reality (a model) and to make it evolve in time. The goal of a simulation model is to reproduce the activities of the simulated system's different entities and therefore to learn something about the behaviors or performance of this system. Several platforms help to form an MAS simulation for predicting the behavior of its agents. In Table 5.2, we present a few platforms which are generally used for simulations and we then discuss some of them.

Product	**Research institute**	**Language**	**Characteristics**
Cormas	Cirad	SmallTalk	*Platform for simulation* For dynamic and use of resource problems Progressive, downloadable platform Several applications on agent society simulation using a resource which can be dynamic and simulated by an MAS Space is a cellular robot 2 types of agents: spatialized and communicating Maximum of 5,000 reactive agents
Geamas	Mas – La Réunion	Java	For modeling and simulation Progressive java platform Several applications on seismic and volcanic phenomena modeling, waste management, study of pelagic resource in the oceanic environment Generic microkernel with which extensions can be paired (learning, self-organization, etc.) Environment is central, the only means of communication Reactive to "strongly cognitive" type agents
Swarm	SWARM Development Group	Objective C, Java	*Platform for simulation* Recursive structure and hierarchy possibilities Many developed applications, particularly in ecology
StarLogo	Media Laboratory, MIT, Cambridge	StarLogo	A programming environment to explore centralized system behaviors

Table 5.2. *Platforms for simulation*

5.5.1.1. *Cormas*

Created by the team GREEN (renewable resource management, environment) within the CIRAD[3] research center (development research center in agronomy, Montpellier), Cormas (common-pool resources and MAS) is an MAS development environment focused on simulations and on the resolution of dynamic process and resource usage problems. Designed as an additional layer over the SmallTalk VisualWorks programming environment, Cormas provides protocol, model and agent libraries which enable the user to develop his own model (possibly relying on existing models) for simulation. Based on the application's needs, Cormas makes it possible to use agents which are very reactive or spatialized (located in space) deliberative, communicating, socially controlled or not, reasoning over time or not, strategists or operational.

Cormas model design consists of three main steps:

– agent, space and communication definitions;

– initialization and simulation control definition;

– definition of the observation of this simulation.

The agents, defined based on classifications during the first phase, can be of several types (agent, located agent, communicating agent, located and communicating agent). Entities must be defined as active or passive for simulation purposes. Passive entities may be simulation objects (whether they are part of the environment or not) or messages exchanged by the simulation's communicating entities.

In order to complete the first phase, the user must define communications, or more precisely, the messages and communications protocols between communicating agents. The user can also visualize the communications between players in the form of a graph similar to the GraphTracer paradigm provided with JDK 1.2.

The definition of space is optional and corresponds to cellular robot development representing the model's spatial dynamics, i.e. definition of the cells and their dynamics. For numerous models, cell dynamics are defined without taking into account the very emblematic neighboring effects of cellular robots.

During the second phase, the different agents and objects must be initialized by instantiating internal variables. It is also necessary to define the control protocol of

3 http://www.cirad.fr/presentation/programmes/espace/cormas/eng/index.shtml.

the simulation progress. Unfortunately, Cormas does not have a pre-implemented agent scheduler and therefore it is necessary to implement a method corresponding to a simulation period or task. However, it is not easy to be abstracted at agent level design by presuming (from an implementation standpoint) that agents operate in parallel as can be the case on other platforms where the scheduler can be abstracted.

The last stage consists of defining the simulation observation. First, we must determine how many times or how long the simulation will last. Secondly, we must determine, with a graphical interface, the general parameters for this space, grid dimension (in number of cells), type of cell surroundings (4-together, 6-together, 8-together), closed or open space. In order to do this, we must determine if the execution will be single-step or continuous and if it is continuous, then how many periods it will last. This corresponds to the number of times the simulation evolution method will be invoked.

If the graphical observation interface is not suitable for observing the model's phenomenon, it is possible to develop another one with VisualWorks such as, for example, progression curves in demography.

5.5.1.2. *Swarm*

Swarm is a generic platform made up of a software library (in Objective C with a Java interface) for the development of agent-based simulations. Its original design was done by Langton[4] (artificial life) and developed at the Santa Fe Institute (New Mexico). The objective is to define a development platform for decentralized "information technology experiments" with discrete events of complex systems. It is a simulation environment of large scale agent societies (in number of agents).

In Swarm, a simulation consists of a group of agents and a calendar indicating what message an agent can send and when. The agent and calendar combination which defines the exchange dynamics constitutes a model in Swarm. This platform is very well known in the artificial life world.

The steps to follow in designing Swarm models are:

– creation of a virtual world: spatial and temporal artificial environment where entities evolve;

– creation of observation agents which will survey the previous virtual world;

– global operation (virtual world + observers) in a time discrete simulation with synchronized clocks.

4 Swarm of the Santa Fe Institute: http://www.swarm.org.

5.5.1.3. *Geamas*

Geamas (generic architecture for multi-agent simulations) is a generic software platform for multi-agent modeling and simulation implemented entirely in Java. It was developed by the MAS2 team at the University of La Réunion.

Geamas software architecture is structured in three modules: a microkernel, a generation environment and a simulation environment. The generic microkernel, called Jaafaar [CAL 99], offers the minimum mechanisms and structures required for implementing MAS, around which a certain number of specialized software extensions (learning, self-organization, assisted design, etc. modules) revolve. Application graphic design is made possible by the generation environment and the simulation environment enables the observation of the simulation progress through a user's graphical interface tool.

5.5.2. *Implementation platforms*

These platforms are used to implement agent architectures designed to make agent development easier. Existing architectures range from simple active objects (for example, Actalk), or from active objects with a communication language such as KQML, to more complex architectures based on conceptual proposals such as mental states.

Implementation platforms can be divided into sub-groups such as design and implementation, or design, implementation and validation as is the case in the study by [GUE 01]. We only present a few in this chapter (Table 5.3) without going into design and validation details.

Product	Research institute	Language	Description
MadKit	Madkit Development Group	Java Scheme, Jess	Multi-agent development tool Programmed agents Concept of role, group, community Messages = information Synchronous motor Interface Free organization
Magique		Java	Empty agents Addition of skills Skill exchange Messages = actions Usable with no interface Optional interface Hierarchical structure
OSACA	UTC Compiègne		Open system for asynchronous cognitive agents Development MAS with few complex cognitive agents
Mask	MAGMA (LEIBNIZ), Grenoble	C, C++, Java, Tcl/Tk libraries	MAS kernel Provide libraries of agents (A), of environment manipulation (E), of interaction (I) and of organization (O) as well as programming tools Cognitive agents
Mercure	LIMSI, AEGIS, LGIS	SmallTalk	FIPA compliant MAS development Commercial Technical agents specified by FIPA (directory facilitator, agent management service, agent communication channel, agent naming service) and specialized agents ACL communications Cognitive agents

DIMA	LIP6 (Paris 6), LERI (Reims)	Based on SmallTalk then Java	Development and implementation of MAS For simulations, problem resolution, control systems with real-time constraints Progressive platform Several applications on assisted ventilation, economic model simulation Separating an agent in perception, deliberation and communication modules Notion of metabehavior 3 types of agents: reactive, cognitive, hybrid
Zeus	British Telecom Labs	Java	Agent building environment
Agent Builder	Reticular Systems, Inc.	Java	Integrated agent and agency development Environment made up of two modules: – Toolkit to define agents, agencies and communications – Run-time system which is a system interpreter
Jack Intelligent Agents	Agent Oriented Software Pty. Ltd.	Jack Agent Language	Agent development environment
dMars	Australian Artificial Intelligence Institute Ltd.	C, C++	Agent-oriented development and implementation environment
Jade	Tilab	Java	Implementation environment

Table 5.3. *Implementation platforms*

5.5.2.1. *Magique*

Magique[5] is a generic MAS development platform designed in Java (JDK 1.x) and intended for distributed problem resolution. Magique is a distributed architecture on a heterogenous network. It is basically a model of agent organization which proposes a hierarchical organization and uses reactive and proactive agents.

5 http://www.lifl.fr/SMAC/projets/magique.

Magique's structure makes it possible to propose a mechanism for distributing skills between agents to facilitate the development of a distributed system. Magique exists in the form of a Java API for the development of hierarchical MAS. The MAS designer can rely on the functionalities provided by Magique such as communication between agents, application distribution, dynamic evolution, etc. to develop the necessary application skills [MAT 01].

There are several multi-agent[6] applications which could be developed by using Magique, among which:

– a system of auction sale with purchasing agents with different strategies and an auctioneer agent;

– territory routing by the exploring agents;

– multiplayer games with an agent referee-scheduler and player agents, etc.

Agents in Magique

Magique enables the distribution of skills among agents and then organizes these agents into a hierarchy [SEC 03]. An agent is an entity with a specific number of skills fulfilling its role in a multi-agent application. Because of exchanges with other agents, an agent's skills can dynamically evolve throughout its life cycle, thus implying that the roles that it can play and its status can evolve within the MAS. An agent is dynamically built from a basic "empty" agent, either by enriching or acquiring its skills. In implementation, a skill can be perceived as a software component combining a coherent set of functionalities and the skills can be developed independently of any agent and reused in different contexts. Adding new skills to an MAS application is easy. An agent's life cycle in Magique consists of executing the following algorithm:

– when receiving a request, expressed in the form of a method name and a parameter list, the agent checks if its class defines the method requested;

– if the agent possesses this method, it invokes and returns the result to the agent which emitted the request;

– if not, it searches within its team to see if one of its subordinates can fulfill the request;

– if that is the case, it asks it to process the request, otherwise it delegates the request's management to the next agent up in the hierarchy.

6 Different examples http://www.lifl.fr/magique, Toy Problems section.

Magique: organizational model

The development of an MAS with Magique initially consists of defining a set of Java classifications characterizing the various agent classifications. In the second phase, the organization of these agents is defined through a hierarchy. Then, agents overloading their *action()* method initiate interactions within the system. This MAS development approach is very similar to traditional software development where after an analysis phase, the system is implemented and tested.

Agent organization can be made to evolve dynamically if a reorganization of the MAS structure is justified during use. In Magique, this process operates on several levels:

– individual: an agent can acquire or discard skills following an exchange with other agents;

– relational: familiar connections can be created when preferred relations appear between two agents of a hierarchy;

– organizational/architectural: agents can be created or destroyed in order to adapt the system to certain constraints. Two examples among others are:

- an agent can be flooded by a surge of requests for one of its skills. It can then decide to create a team of agents which will have this skill and redirect all or part of the requests to them,

- an agent must be able to leave the system temporarily and find its place later; the messages intended for this agent must then be put on standby.

5.5.2.2. *Zeus*

Designed by British Telecom, Zeus is a complete environment based on the studies of the FIPA, using a methodology called role modeling to develop collaborative applications. Its concepts rely on notions of agents, goals, tasks (that the agent must execute to achieve its goal) and facts (what the agent considers real). Zeus[7] provides an agent development environment with the help of a set of Java libraries, which developers can reuse to create their agents. Zeus also proposes a group of utility agents (naming and facilitation server) to facilitate the search for agents.

Zeus includes a set of components written in Java which can be grouped in three operation categories (or libraries): *agent component library, agent building tool, agent utilities* containing a name server, a facilitator and an agent viewer. It is one of

7 Zeus Agent Building Toolkit: http://www.labs.bt.com/projects/agents/zeus/index.htm.

the most comprehensive tools available. The various development stages are completed through several editors: ontology, task description, organization, agent definition, coordination, facts and variables as well as constraints.

Zeus agents

The Zeus agent architecture is similar to most collaborative agent architectures. Agents have the following characteristics:

– an agent has a *definition*;

– it belongs to an *organization*;

– it *perceives* its environment;

– it *modifies* its environment;

– it has a life *cycle*;

– it relies on an interaction *protocol*.

A Zeus agent is made up of three layers:

– the definition layer, which contains the reasoning capabilities and learning algorithms. The agent is seen as an autonomous entity able to reason in terms of beliefs, resources and preferences;

– the organizational layer, which contains the agent's knowledge base, as well as the relations between agents;

– the coordination layer, where communication modes are decided between agents, protocols, coordination and other interaction mechanisms.

The Zeus architecture lends itself well to planning or scheduling applications, which makes it particularly interesting for simulation or monitoring application development. However, each agent functions with a JVM (Java virtual machine), which quickly becomes difficult to manage when the number of agents in the system is too large.

5.5.2.3. *Jade*

Jade[8] (*Java agent development framework*) is a tool which meets FIPA standards. It is a middleware written in Java and compliant with FIPA specifications. This environment simplifies agent development by providing basic FIPA-defined services, as well as a set of deployment tools. The tool has three main modules

8 http://sharon.cselt.it/projects/jade.

(necessary for PIPA standards): DF (director facilitator) provides the platform with a service of yellow pages, ACC (agent channel communication) manages the communication between agents, AMS (agent management system) monitors agent registration, their authentication, their access and system use. The agents communicate with FIPA ACL language. An editor is available for agent registration and management. No other interface is available for the development or implementation and because of this the implementation is arduous. It requires good knowledge of the classes and various services offered.

A Jade platform can be distributed over a set of machines and configured remotely. However, the system configuration can be modified dynamically because of an agent migration mechanism within the same platform.

5.5.2.4. *AgentBuilder*

The AgentBuilder[9] platform is a complete development environment. Object oriented modeling constitutes the basis for system design to which we add an "ontology" part. It is an implementation similar to the Agent-0 language. Agents are defined in Radl (Reticular Agent language definition) language which makes it possible to define agent behavior rules. KQML is used as communication language between agents and rules are launched according to certain conditions and are associated with actions. The conditions relate to messages received by the agent, whereas the actions correspond to the use of the Java method. System execution is done through the AgentBuilder execution engine. We can, however, create ".class" files and execute them on a standard JVM (Java virtual machine). It is also possible to describe protocols defining accepted and transmitted messages. AgentBuilder is probably the most successful commercial platform. AgentBuilder is a complex tool which requires a steep learning curve and a solid MAS knowledge in order to be used in an efficient manner. It is limited by its scalability, deployment and reuse.

5.5.2.5. *Jack Intelligent Agents*

Jack Intelligent Agents[10] is an environment made up of a project managing editor, a JAL (jack agent language) Java extension programming language and a compiler. It is intended for the development of BDI-type agents, while providing an object oriented API. The language proposed to the developer is similar to the logical programming and can be compiled to transform it into Java classifications. The generated classes extend classifications provided by the Jack core API: agent class, event class or plan class. The Jack core also controls task competition between the agents, event reaction and the communication infrastructure. The project manager is

9 http://www.agentbuilder.com.
10 http://www.agent-software.com.au.

an interface including a text editor where the system implementation is done. System compilation (from JAL to Java) and execution are also done within this interface.

5.5.2.6. *MadKit*

The MadKit[11] platform developed at the University of Montpellier II is based on the Aalaadin methodology [FER 98] and the concept of agent, group and role. MadKit provides an API for the development of agents by specifying an abstract class of agent. Each agent can play various roles within different groups. The agents are launched by the MadKit node. The tool provides an editor enabling the deployment and management of MAS (a-box). Management done with the help of this editor offers several interesting possibilities. It is possible to exchange messages directly with an agent or a group of agents. This platform is especially interesting for the organizational approach which it emphasizes during the analysis and design of an MAS. The tool also offers a utility for carrying out the simulations.

5.5.2.7. *DIMA*

DIMA (development and implementation of MAS) is an MAS development environment created within Guessoum's[12] thesis. DIMA proposes reactive, cognitive, hybrid, autonomous and adaptable agents. The first version of DIMA was implemented by using the Smalltalk-80 language and was then ported to Java. It provides basic layout libraries for building different agent models. The communication languages between agents are based on KQML and FIPA ACL languages. The DIMA platform is intended for simulations and problem resolution. It was used to develop several real applications like assisted ventilation [DOJ 97], economic model simulation [DUR 00], simulation of electric market, etc.

5.5.3. *Mobility platforms*

To design and implement a mobile agent platform, three approaches are observed. The first consists of finding a programming language which includes instructions for mobile agents. The second is to implement the mobile agent system as operating system extensions and the third approach develops the platform as a specialized application which runs on an operating system [SAM 03]. Most systems use the latter approach, which is often summarized in a collection of Java libraries such as Travelling, Aglet, Concordia, Mole and Odyssey. As an example, we present in Table 5.4 some representative platforms of mobile agent systems.

---

11 http://www.madkit.org.

12 http://www-poleia.lip6.fr/~guessoum/dima.html.

Product	Research institute	Language	Description
Aglets	IBM Japan	Java	Mobile agents
Concordia	Mitsubishi Electric	Java	Mobile agents
Grasshopper	IKV++	Java	Mobile agents
GenA	CRIM	GenA	Mobile agent platform

Table 5.4. *Mobility platforms*

5.5.3.1. *Grasshopper*

Grasshopper[13] is a platform providing the management of agent migration and communication continuity during these migrations. Services provided by the platform are the management of agent life cycle, directory management and migration.

5.5.3.2. *Aglets*

Aglets is a platform developed by IBM Japan to define a mobile agent development framework [SAM 03]. In this context, the platform deals with services which are necessary to agent migration and communication management. An aglet[14] is a Java class extending a class provided by the node and redefining the functions to activate during agent serialization/deserialization. The architecture of an aglet is very similar to a Java applet architecture and the system has its own protocol to transfer aglets between hosts: the aglet transfer protocol.

5.6. Conclusion

Few attempts have been devoted to the standardization of MAS platforms as Ricordel [RIC 00] discusses. Only a few research groups and manufacturers (OMG, FIPA, KAOS, General Magic Group) have begun this standardization process. The second observation is that there is no complete multi-agent platform or environment. A good environment must offer user assistance as is the case in the Mercure platform, it must be open and even self-organized as in MACE, integrate heterogenity as in MASK and DIMA, provide follow-up tools such as MAST or simulation tools as in

---

13 Grasshopper-2 Agent Development Platform: http://www.grasshopper.de.
14 http://www.aglets.org or http://www.trl.ibm.com/aglets.

Cormas, validate a system, even partially, with respect to the required specifications and offer deployment capabilities as in Vulcano.

The objective of this chapter was to give a short description of a few MAS platforms divided into three categories according to their fields of application: simulation, implementation and mobility. It is important to evaluate and compare the technical aspects of MAS development tools as there are few evaluation studies published until now. The summary presented in this chapter indirectly points out the strengths and weaknesses of some of the platforms which were considered significant in a network context.

5.7. Bibliography

[BOI 01] BOISSIER O., Cours SMA-DEA-CCSA Multi-Agent systems, MAS Platforms, SMA/SIMMO, Ecole des Mines de Saint-Etienne, December 5 2001.

[BRA 96] BRADSHAW J., "KAoS: An Open Agent Architecture Supporting Reuse, Interoperability, and Extensibility", Research and Technology, Boeing Information and Support Services, Seattle, 1996.

[CAL 99] CALDERONI S., GEAMAS, PhD Thesis, University of La Réunion, November 1999.

[DOJ 97] DOJAT M., PACHET F., GUESSOUM Z., TOUCHARD D., HARF A., BROCHARD L., "NeoGanesh: A Working System for the Automated Control of Assisted Ventilation in ICUs", *Artificial Intelligence in Medicine*, vol. 11, no. 2, September-November 1997.

[DUR 00] DURAND R., GUESSOUM Z., "Competence systemics and survival, simulation and empirical analysis", *Competence 2000 Conference*, Helsinki, Finland, p. 10-14, June 2000.

[FER 89] FERBER J., GUTKNECHT O., "Aalaadin: a meta-model for the analysis and design of organizations in multi-agent systems", *ICMAS'98*, p. 128-135, Paris, 1998.

[GAR 02] GARNEAU T., "Programmation orientée-agent: évaluation comparative d'outils et environnements", *Systèmes multi-agents et systèmes complexes, JFIADSMA*, Lille, France, 2002.

[GUE 01] GUESSOUM Z., Chapter 5 in Jean-Pierre Briot and Yves Demazeau (ed.), *Principes et architecture des systèmes multi-agents*, 2001.

[MAR 02] MARIETTO M.B., DAVID N., SICHMAN J.S., COELHO H., Requirements Analysis of Agent-Based Simulation Platforms, Technical Report, 2002.

[MAT 01] MATHIEU P., ROUTIER J.C., "Une contribution du multi-agent aux applications de travail coopératif", *TSI*, Hermes Science Publications, Special teleapplication edition, 2001.

[NUN 02] NUNO D., MARIETTO M.B., SICHMAN J.S., COELHO H., *Requirements Analysis of Multi-Agent-Bases Simulation Platforms: State of the Art and New Prospects*, 2002.

[PIE 03] PIERRE S., *Piranhas 2, Concepts et réalisation*, January 24 2003.

[RIC 00] RICORDEL P.M., DEMAZEAU Y., "From Analysis to Deployment: a MultiAgent Platform Survey", in A. Ominici, R. Tolksdorf and F. Zambonelli (ed.), *1st International Workshop on Engineering Societies in the Agents World (ESAW), ECAI'2000*, Berlin, August 2000.

[SAB 01] SABAS A., "Systèmes multi-agents: une analyse comparative des méthodologies de développement, vers la convergence des méthodologies de développement et la standardisation des plates-formes SMA", University du Québec à Trois-Rivières, October 2001.

[SAM 03] SAMUEL P., *Réseaux et systèmes informatiques mobiles. Fondements, architectures et applications*, International Press Polytechnique, 2003.

[SEC 03] SECQ Y., "RIO: rôles, interactions et organisations, une méthodologie pour les systèmes multi-agents ouverts", University of Lille, Supported thesis report, 2 December 2003.

[SIM 01] SIMULATION OF COGNITIVE AGENTS, "Requirements Analysis of Multi-Agent Based Simulation Platforms, Simulation of Cognitive Agents", Technical report, 2001.

[WHI 96] WHITE J., General Magic, Mobile Agents White Paper, General Magic, 1996.

Chapter 6

Behavioral Modeling and Multi-agent Simulation

6.1. Introduction

Telecommunications networks must increasingly integrate dynamicity because of the heterogenous nature of the data processed. In fact, within the same network, packets will transmit (voice, video, etc.) with very different quality of service and processing requirements. Moreover, we are experiencing an expansion of new networks such as mobile, active and policy-based networks. These networks are striving to offer their users more services, thus generating an internal processing complexity at router and terminal level. By dynamically adapting management mechanisms already deployed, they are trying to ensure a better network resource use in relation to current conditions. Intelligent agents are often used to achieve this adaptability within the network.

The new types of networks and the methods used to ensure their management control will introduce more adaptability and dynamicity into the network. However, they will increase network modeling and simulation complexity. In fact, analytical methods cannot represent future network operation and behavior because they only represent set, fixed situations. Simulations are then the solution offered to study the behavior of these networks.

Mainly based on static approaches such as queues, the current network simulation techniques are not able to take into account the excessively large number

---

Chapter written by Leila MERGHEM-BOULAHIA.

of states that a dynamic network can have. It is therefore necessary to find a new simulation method which can represent a dynamic network and study its performances in order to provision it.

Multi-agent modeling and simulation seem to be the most popular approaches. Multi-agent systems have advantages giving them an interesting predisposition for the representation of dynamic environments. They are used in several sectors and have proven their reliability and capacity to represent, understand, explain and even discover new phenomena. Due to their main attributes of autonomy, sociability, communication, cooperation, proactivity, learning, etc. [WOO 02, BAR 01], agents are able to model a dynamic network environment.

The purpose of this chapter is to show the advantage of the multi-agent approach, particularly that based on the concept of behavior by describing its principles and a few of its contributions. The remainder of this chapter is organized as follows: we initially describe three main telecommunications network modeling approaches by defining the principles, advantages and limits of each approach. The inability of these traditional approaches to consider the highly dynamic nature of new networks has justified resorting to a new approach: multi-agent systems. Section 6.3 will present multi-agent modeling and simulation by emphasizing the characteristics, steps, advantages and limits of this approach. Section 6.4 describes the principles of behavioral modeling as well as other studies constituting a source of inspiration for our contribution which is presented in section 6.5 and consists of behavioral modeling of a network node organized in two levels: basic behaviors and metabehaviors. All aspects related to this model will be largely studied. Section 6.6 concludes this chapter.

6.2. Traditional network modeling and simulation approaches

In order to study the operation of a telecommunications network, two approaches can be used: the analytical approach and simulation. Because of their significant dynamic, new networks function in a non-deterministic way. It is therefore not possible to rely on an analytical approach for network representation and simulation is the only suitable way to achieve this task. Before moving on to simulation, a model representing the main system characteristics should be provided. For this purpose, we will focus on the three main approaches used to model the functionalities of a telecommunications network: queues, Petri nets and process algebra. We will briefly describe the characteristics of these approaches.

6.2.1. *Queuing theory*

The purpose of queuing theory is to model systems where "clients" need a "resource" for the execution of their activity. Since this resource is being solicited by several clients, it is not always free. Hence, the advantage of the queue concept where clients are put on a queue until the resource is available. The queuing theory was used in fields such as information technology systems (process management), production systems (inventory control), communication networks (traffic management), etc. [BAY 00].

In order to specify a queue, it is necessary to characterize the client arrival process, time of service as well as service discipline and structure of the queue. The description of queues was standardized by Kendall's observations which took into account interarrival and service distribution, number of servers, user population and service discipline. Queuing theory applies well to deterministic networks whose structure is set [PUJ 02].

6.2.2. *Modeling by Petri nets*

A Petri net is a specific type of directed graph, with an initial state called initial marking, M0. Two different types of nodes can be observed: places and transitions. Places are the equivalent of conditions and transitions are the equivalent of events. In order to explain the dynamic behaviors of a system, the Petri net that models it changes states due to the activation of possible transitions.

Petri nets can simply be used in any field or system which can be graphically described and which requires a way of representing parallel and concurrent activities [MUR 89]. However, careful consideration must be given to the compromise between the model's general nature and analysis capacity. Indeed, a major weakness in Petri nets is their complexity. The models based on Petri nets tend to become very large even for the analysis of a medium-sized system.

6.2.3. *Modeling by process algebra*

Stochastic process algebra is a subject that has been relatively well studied, with a significant number of available tools and techniques such as PEPA, TIPP [HER 01], etc. [BOW 01]. The most important technique is called PEPA (*performance evaluation process algebra*), which is a tight language providing vital and simple tools to describe the system.

In a PEPA [GIL 94, HIL 96], a system is described as being the interaction of components committed to activities, individually or collectively. The components correspond to identifiable system sub-structures or to roles in system behavior. They represent active units within the system. A queue, for example, can be regarded as consisting of an arrival component and a service component interacting to shape the queue's behavior. A component can be atomic or made up of other components.

PEPAs propose several operators representing cooperation, choice (competition), etc. PEPA characteristics such as composition, economy and the existence of a formal definition make them well adapted to complex systems with parallel activities. However, state space explosion is the major disadvantage of this approach. Another disadvantage of PEPA, even if it also contributes to their strength, is the restriction on exponential distributions, whereas in reality the applications can have components which are uniformly distributed or constant [GIL 94].

6.2.4. *Limits*

Simulating a highly dynamic system is not an easy task. The model developed must reflect the target system as accurately as possible without it becoming too complex and thus, difficult to analyze. By nature, new telecommunications networks are very dynamic and not easily modeled by traditional methods. Queuing theory is largely used because it is suitable for deterministic networks. However, it cannot model a network in which node decisions depend on information coming from other nodes or node experience, or even from other parameters like time of day, etc.

Petri nets make dynamic systems modeling possible. However, the modeled system must have a limited number of states so that it does not become too complex and difficult to analyze. The same assessment applies to the process algebra which, because of their main characteristics such as composition, existence of a formal definition, etc., have an interesting predisposition for representing complex systems. On the other hand, state space explosion is a problem that has also been encountered with this approach.

Since traditional modeling and simulation methods are not able to represent new networks easily, we have chosen another simulation approach. This solution is based on a behavioral multi-agent approach, which has already proven reliable in numerous complex and dynamic domains other than telecommunications networks.

6.3. Multi-agent modeling and simulation

Multi-agent modeling and simulation was used in several fields including road traffic simulation [MOU 98, BAZ 03], social phenomena simulation [BEN 03, VAN 02, CON 98], biological phenomena simulation [DRO 93, DOR 01, EDM 03], etc. In addition, other proposals have been made for the use of agents in the telecommunications networks field in order to solve very specific problems such as topology detection of a dynamic network [ROY 00], routing optimization [SIG 02], fault detection [WHI 99], minimization of call blocking probability in a mobile network [BOD 00], congestion control [FAR 00], etc. To our knowledge, there are no studies which use a multi-agent approach to simulate telecommunications networks. This may have been justified until very recently because traditional approaches were adequate, but now, following the dynamic network growth, it has become necessary to rely on an intelligent and adaptive approach.

6.3.1. *Multi-agent simulation steps*

The multi-agent simulation is generally used to study the operation of a system when mathematical methods or other methods usually used (cellular robots, Petri nets, etc.) are not sufficient anymore. A simulation is built from a model mirroring the most relevant aspects of a target system. The basic sequence of a multi-agent simulation is described below [EDM 03] (Figure 6.1):

– model design: the various system elements as well as their functionalities and interactions with other system components are defined. Each element is then represented by an agent for which we specify functionalities, types of relations and interactions with the other agents, communications and cooperation mechanisms, etc.;

– inference: the agent model designed during the previous step is implemented to obtain an application which runs during a certain amount of time;

– analysis: the resulting process is analyzed with a variety of methods including statistics, the Monte Carlo method, visualization techniques, etc.;

– interpretation: finally, the goal of this exercise is to arrive at conclusions in relation to the target system by interpreting all the deductions related to the behavior exposed during the various MAS executions.

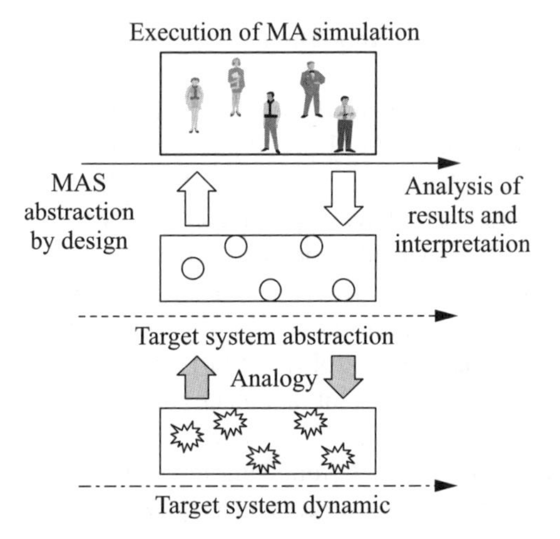

Figure 6.1. *Steps in a multi-agent simulation*

6.3.2. *Contributions*

Multi-agent simulation with its heterogenous agent types (reactive, cognitive, hybrid) makes it possible to consider the nature of the modeled system's components. Logically, it makes more sense for a human being to be represented by an agent which has reasoning and planning capabilities (cognitive agent), whereas for an ant it makes more sense to be represented by an agent which responds to events with simple actions (reactive agent). It is even possible to abstract the nature of the agents represented by emphasizing their organization first. This is the case, for example, of the Aalaadin model [FER 03] which is based on the concepts of agent, group and role. In this model, we initially observe groups of agents designed to execute a specific number of goals and the nature of the agent depends on its role within the group. It is thus possible for an agent belonging to several groups to have cognitive aspects in a given group (president of a company) while being considered reactive in another group (spectator in a concert).

The multi-agent approach makes it possible to choose the level of abstraction on which we want to be placed according to our simulation objectives, such as the level of detail desired and data available on the system. We can thus choose to study the behavior of one ant or a whole colony of ants. In the network field, we can study the operation of each router or the router of a given domain, etc. By having the possibility of choosing a level of abstraction, it becomes possible to verify the relevance of the level usually selected. This is possible by comparing the system comprehension and/or representation gains obtained with each selected level of abstraction. In general, we match a real entity of the target system to an agent representing its characteristics and behaviors. However, it is possible to consider agents which do not necessarily represent a real entity but rather a group of these entities. In fact, in the same multi-agent simulation, it is possible to have several time and space scales by representing, in one system, agents known as "microscopic" and others known as "macroscopic". Macroscopic agents are generally the result of a grouping of microscopic agents. The works of Servat [SER 00] in the Rivage project are a very good example of such a methodology. The study examines the runoff dynamic, infiltration and erosion. Microscopic agents represent water drops which, according to their physical environment, can gather and constitute agent pools or streams. The dissolution of these agents can occur if the water drops constituting them change directions.

Such a wealth of models is likely to represent a large range of completely different systems. We should note that for now, the multi-agent approach is only used when traditionally used approaches become inadequate. This implies that it is not always easy to validate completed simulations. The total validation of the operational solution (multi-agent simulation) becomes impossible without having any convincing theoretical results from the target model [MEU 01]. The only validation that is possible to achieve consists of checking the relevance of the model and of the simulation operation with the target field's expert.

6.4. Behavioral modeling

6.4.1. *Principle*

Behavioral modeling is multi-agent modeling that defines the behavioral functionalities of a system's component. Each component has a set of behaviors representing the tasks that it accomplishes within the system. The interaction between the various behaviors, either within the same component or between various components, generates the process for the complete system and it is possible for us to observe it through the inference step (section 6.3.1).

The behavior reflects the activities of an entity and needs to be represented in the simulation. These activities depend on the simulated environment. In this way, a driver in a road traffic simulation will not have the same behavior (and therefore the same actions) as an ant in an ant colony simulation, or as a router in a network simulation. Behavioral modeling was mainly used to represent the operation of a multirobot system as well as social or socio-economic phenomena. The works of

Matarić and Drogoul and their teams in these domains are very good examples [MAT 94, NIC 02, DRO 93, BEN 03].

6.4.2. *Contributions*

Matarić [MAT 94] defines behavior as being a law of control which groups constraints in order to reach or maintain a given goal. She supposes that, for each field, there is a set of what she calls "*basic behaviors*" which are required to generate other behaviors, while being a minimum unit needed by the agent to achieve its goals. Choosing this set of basic behaviors wisely is vital. Behaviors are defined and implemented in the collective robotics environment. A behavior involves a group of agents rather than a single one (the *following* behavior involves at least two agents: an agent at the head and another following it). The generation of compound behaviors requires the application of combination operators whose priorities are well understood and which produces the desired composite output behavior.

In any behavior-based system, the arbitration between the various behaviors (co-ordination between the activities of several input behaviors in order to produce the desired output behavior) constitutes a real challenge. The conditions of mutual exclusion between behaviors are effective for arbitration in a one-behavior system at a given time. However, in a more complex system such as collective robotics, several behaviors can contribute to the output behavior. Another level of arbitration is therefore necessary and will implement a type of summation (behaviors are executed at the same time) or switching (behaviors are carried out sequentially) of activated behaviors.

A model based on "*primitive behaviors*" and "*abstract behaviors*" is proposed by [NIC 02]. The original contribution to this model is the notion of abstract behaviors which is actually only the association of activation conditions of behavior and its effects without specifying its internal procedure. An abstract behavior is then simply the implicit specification of the execution conditions of the behavior and its effects. Behaviors executing the work that will ultimately give the desired results under specific conditions are called primitive behaviors and can include one behavior or a whole collection working either sequentially or in parallel. The authors expand the abstraction to a higher level by defining what they call behavior networks where nodes are abstract behaviors and the links between them represent the relation between their pre-conditions and post-conditions.

Drogoul [DRO 93] proposed a behavioral model to describe the operation of a colony of ants. In this model, “the behavior of an ant is characterized by a set of independent tasks, made up of a sequence of basic controlling behaviors called primitive. Each of these tasks is exclusive and its launch comes from an external or internal stimulation, made real by variable force stimuli, that combines with a pre-existing motivation expressed under the double aspect of inhibiting threshold and weight, a reflection of the previous animal experiment in this task” [DRO 93]. In the same spirit, Bensaid [BEN 03] defines the behavioral model of a consumer in a competitive market. A consumer has a set of primitive behaviors (imitation, opportunism, etc.) and is influenced by past experience and external stimuli, which enable him to decide the action to take in relation to a given brand.

These studies have led to several behavioral models, each with its own characteristics [MER 03]. However, the common denominator which seems most important to us is the distinction between two categories of behaviors: simple behaviors called basic behaviors or behavioral primitives, etc. depending on the model, and other more complex behaviors. This second category of behaviors is the result of the combination of simple behaviors by executing them sequentially or in parallel. The purpose of this combination is to efficiently use the system’s simple behaviors to complete more complex tasks requiring the execution of several simple behaviors.

Inspired by these various models, we also proposed a two-level behavioral model of a network node described in the following section.

6.5. Two-level behavioral model of a network node

6.5.1. *Introduction*

Given the inability of traditional approaches to take the increased dynamic of new QoS networks into account, the use of the behavioral multi-agent approach is obviously justified. Therefore, we have proposed a behavioral modeling of network nodes to represent their process. We are conscious of the existence of another important aspect which influences the operation of a network and is called user profile. The behavioral modeling of user profiles is a subject that is not developed here but will be in future works. We will only address the behavioral model of network nodes.

To complete its various tasks, the node uses several mechanisms which are represented in our model by a set of basic behaviors managed by a metabehavior. In the proposed model, we observe two behavioral levels within a network node (Figure 6.2). These two levels represent the two decision levels in a node:

– basic behavior level (level 0): this level includes the basic behaviors which could achieve control tasks such as scheduling, queue management, etc. The basic behavior chosen by the metabehavior to ensure a given control continues to carry out its actions as long as the metabehavior has not designated another behavior to complete the task at hand. This new basic behavior is selected by the metabehavior because it is considered more relevant to the current state of the node and of its neighbors than the previous one;

– metabehavior level (level 1): this level monitors, observes and controls the basic behaviors of level 0. In fact, it chooses the basic behaviors to be activated. In Figure 6.2, we can see that the metabehavior can choose which of the four basic behaviors to activate. In this example, basic behaviors A, B and C are activated and basic behavior D is inhibited.

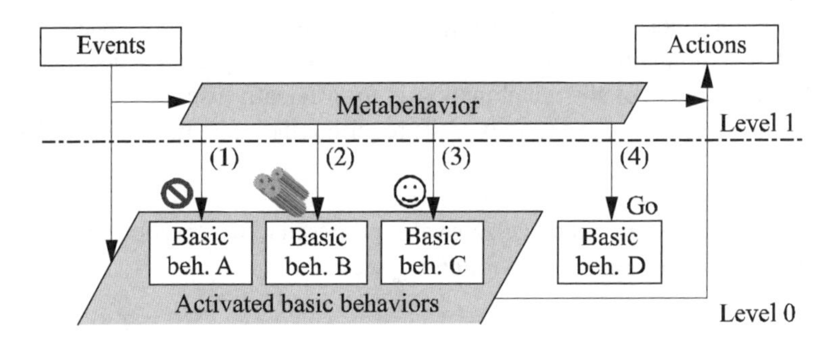

Figure 6.2. *Operations on basic behaviors*

A node owns a set of metabehaviors (metabehavior A, etc., metabehavior Z) which must provide it with an adaptive, consistent and optimal operation. In order to complete a determined control task, each metabehavior calls for the parallel or sequential execution of a set of basic behaviors. In Figure 6.3, for example, metabehavior A is responsible for launching the execution of basic behaviors a_1, a_2..., a_n.

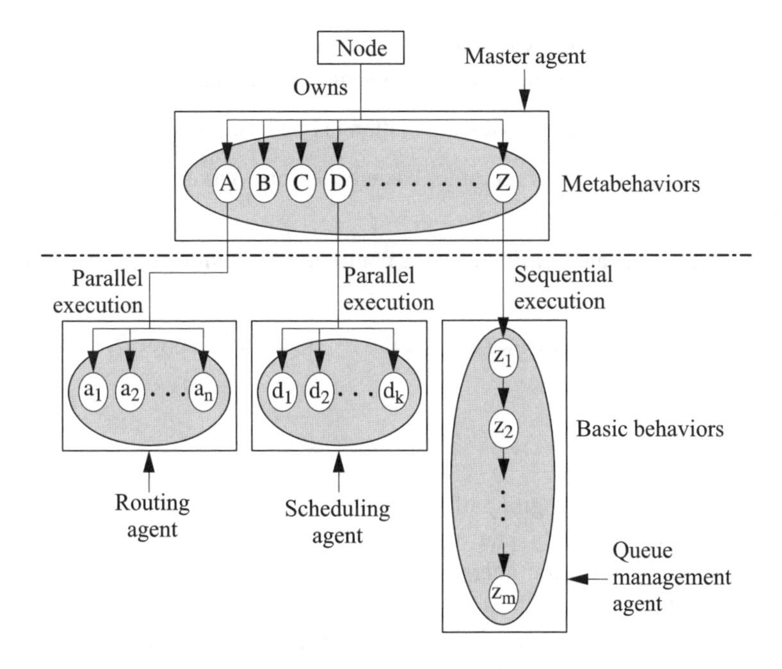

Figure 6.3. *Behavioral model of a node*

Our approach consists of representing the different network nodes by agents which optimize specific network parameters, by adopting appropriate basic behaviors for the current state of the received traffic and surrounding nodes. The dynamic of the network is therefore the result of each node's interactions with the traffic it receives by considering its current state and the behaviors that it exhibits.

6.5.2. *Role of the two behavioral levels*

The role of the basic behavior level is to model the mechanisms which enable the node to correctly manage the data it receives with a specific QoS guarantee. Some basic behaviors model mechanisms dealing with queue management whereas others model mechanisms for packet routing, etc. On the other hand, the role of the metabehavior consists of modeling the dynamic and adaptive QoS mechanisms. It represents the mechanism which monitors the current state of the node and decides which basic behaviors are most appropriate for the observed state. Therefore, metabehavior constantly decides which basic behaviors to inhibit and which ones to activate, and with what activation parameters (Figure 6.2).

Because of its metabehaviors, a node can change the activated basic behaviors following specific internal and/or external events. These changes can consist of:

– inhibiting certain basic behaviors (action (1) on basic behavior A, Figure 6.2);

– changing the parameters of other basic behaviors (action (2) on basic behavior B, Figure 6.2);

– letting the activated basic behavior continue its activities (action (3) on basic behavior C, Figure 6.2);

– activating other basic behaviors (action (4) on basic behavior D, Figure 6.2).

6.5.3. *Agents*

A node is made up of a group of agents that are classified in two categories. The first category contains only one agent called master agent, whereas the second category can contain several agents such as queue management agent, scheduling agent, etc. (Figure 6.3). Each agent in this category is responsible for a specific operation provided by the basic behaviors that it produces (the queue management agent, for example, has basic behaviors for managing the queue). The master agent monitors the remaining agents. In fact, because of its metabehaviors, it is responsible for the basic behavior selection of the various agents that it manages depending on the current situation and occurring events.

Several questions can be asked: why so many agents? Would it not have been simpler to only have one agent dealing with all the node tasks? The answer to this last question is obviously yes: it is entirely possible to have only one agent, but we must be conscious of the consequences that such a choice might generate. With this solution, an agent's operation would be complex and updates and expansions difficult. By using several agents which have a simple operation on one side, and one agent taking time to observe the system's evolution and intervening only at specific time intervals on the other side, we think we can arrive at a simple, efficient and easily expandable representation of new network dynamics.

6.5.4. *Model of two behavioral levels*

6.5.4.1. *Basic behavior model*

In our model, a basic behavior has three main parts: its inputs (activation and stimuli), its internal part, and its outputs (actions). The *activation* input reflects the state of activation of the basic behavior and the input *stimuli* groups the events triggered by this basic behavior. These stimuli can be internal or external. The *actions* output groups all the actions which influence the environment (packets, surrounding nodes, etc.) or the node's inherent components (queue, routing table, etc.) which are likely to modify the system's components function. The action launched by the basic behavior can have immediate consequences such as packet reject, or indirect effects (deferred). An example of indirect effects can appear after a congested node sends a message to its neighbors. The immediate consequence is sending the message, whereas the indirect effect can be a modification of the basic behavior handling routing at the level of the surrounding nodes. The internal part of a basic behavior contains its entire logic.

6.5.4.2. *Metabehavior model*

The node has a range of basic behaviors ensuring a given control. Each one of these basic behaviors can guarantee good performances under certain network conditions. Nonetheless, no basic behavior will always give the best performance whatever the network configuration, number of customers, nature of traffic, etc. because these basic behaviors only process a small part of a defined control task. They are basic and because of this they do not achieve anything very complex such as the adaptation of their conditions and activation parameters. It is now necessary to adapt node management mechanisms to the new conditions. Since basic behaviors, by definition, do not have auto-configuration capabilities, this task is then assigned to metabehaviors.

The metabehavior chooses which actions to take by consulting the current state of the system (state of surrounding nodes, loss ratio, queue occupancy rate, etc.) and its rules. It acts on the basic behavior that it monitors in two ways:

– by changing the value of the basic behavior's "activation" input. Based on the data collected and its rules, the metabehavior may assume that it would be more useful to activate (or inhibit) this basic behavior;

– by modifying the internal procedure of the basic behavior in question. The basic behavior depends on this modification which appears during parameters update, such as MinTh and MaxTh thresholds for basic "careful" behavior (Chapter 7, section 7.2.1). It is more logical that these thresholds are higher when a large *best effort* packet loss is reported.

Two design choices are possible:

– a global metabehavior. In this case, the metabehavior checks all the control tasks of a node. For a node, the metabehavior handles the management of the queue, scheduling, routing, etc. all at the same time. The metabehavior will have rules for managing the queue, others for routing, etc. This approach is simple because the designer does not need to specify with precision what the various system tasks are, but may experience expansion and update problems. Another disadvantage with this approach is the arbitration difficulty between activated basic behavior actions (section 6.5.5.2). In fact, it is easier to define an arbitration policy between actions related to one control task rather than when the actions of several tasks are mixed;

– dedicated metabehaviors. In this case there will be several metabehaviors, each one dedicated to a specific task. A metabehavior handling queue management, for example, does not review routing or admission control rules, etc. The advantages of this solution are directly inherited from those of the modular design approach such as, for example, reuse, ease of expansion and update, etc. The arbitration between the different basic behavior actions to ensure node consistency is achieved more easily.

Our choice focuses on the second approach because of its advantages. Since the node control tasks largely depend on the involved network's context, it is important to specify that the metabehaviors within a node will not always be the same. A given architecture can involve only routing and a simplified queue management if all the packets are from the same traffic class (best effort, for example). In this case, the only metabehaviors to use are routing and queue management metabehaviors and, depending on the context, they will activate the necessary basic behaviors. A DiffServ type network [BLA 98] on the other hand requires other metabehaviors such as admission control, scheduling, etc.

6.5.5. *Ensuring adaptability*

To complete a given control task, the metabehavior has a library of basic behaviors among which it must choose the most relevant by referring to its rules and the current environment conditions. The metabehavior can execute its basic behaviors in two different ways: parallel or sequential execution.

Adaptability is ensured at node level because of the continuous change of activated basic behaviors to accurately and effectively respond to current conditions. These changes are the result of the execution of metabehavior rules.

6.5.5.1. *Metabehavior rules*

The model of a rule is illustrated in Figure 6.4. An event launching a rule can correspond to the arrival of a message, the reaching of a threshold, the value of a parameter or the value of the clock [MGP 03].

Rule:: = <Events>; <Actions>
Event:: = <message>|<Parameter_value>|<Clock_value>|<threshold>
Message:: = <simple_message>|<normalized_message>
Simple_message:: = <Source>; <Destination>; <Msg_Type>
Normalized_message:: = <KQML_message>|<ACL_message>
Parameter_value:: = <attribute>;<operator>;<value>
Clock_value:: = <current_agent.clock>;<operator>;<value>

Figure 6.4. *Model of a rule*

The role of metabehavior rules is to manage a specific number of basic behaviors in order to ensure the best possible node operation and to prevent inconsistent decisions within the same entity. They make it possible to choose the basic behaviors to activate, those to inhibit and the appropriate activation parameters. Each metabehavior has a set of inherent metabehavior rules and some examples are given in section 7.3 in Chapter 7.

6.5.5.2. *Arbitration between basic behavior decisions*

Since activated basic behaviors are able to make different decisions, the metabehavior must choose only one action to launch in order to keep router consistency. Indeed, it would not be logical to apply two completely contradictory actions to the same packet: rejection and acceptance, for example. Several heuristics can be considered:

– choosing the most restrictive action (packet rejection in the case of queue management, for example);

– choosing the intermediate action (class marking or deterioration: a *premium* class packet becomes *olympic* and an *olympic* class packet becomes *best effort*);

– choosing the least restrictive action (acceptance);

– choosing the action determined by the majority of basic behaviors.

6.5.5.3. *Relations between basic behaviors*

The main goal of the relation study between basic behaviors is to keep node consistency by preventing scenarios where basic behaviors launch simultaneous contradictory actions. These relations are:

– contradiction: two main behaviors are considered contradictory when, under the same conditions and with the same environment knowledge, they launch contradictory actions. Consistency is ensured by centralizing the main actions of the activated basic behaviors that deal with the same control task because of metabehavior rules (section 6.5.5.2);

– indifference: two basic behaviors are considered indifferent if they complete different node control tasks. A relation of indifference exists, for example, between a basic behavior dealing with routing and another dedicated to queue management;

– complementarity: two basic behaviors are said to be complementary if the existence of one is reinforced by the existence of the other. We can observe a complementarity between certain basic behaviors managed by the same metabehavior. For example, the basic "careful" behavior is reinforced by the basic "thrifty" behavior because their coexistence makes it possible to keep only priority packets (because of basic careful behavior) and useful packets (because of basic thrifty behavior) in the queue (Chapter 7, section 7.2.1). Complementarity can also exist between the metabehaviors of one node. Admission control metabehavior can supplement the work of queue management metabehavior – when it is flooded by the large quantity of traffic it receives – by activating admission control basic behaviors that refuse connections which could intensify queue congestion. This complementarity can also appear on a larger scale and consequently lead to a cooperation between different routers when it includes a router and its close neighborhood. Careful basic behavior (Chapter 7, section 7.2.1) activated at router level and where the queue might soon become saturated (if not already) shows better performances when its neighbors are "disloyal" than when they are "conservative".

6.6. Perspectives and conclusion

The understanding and anticipation of the functionalities of a complex and dynamic system usually requires simulation. In this chapter, we have studied multi-agent simulation by describing its various steps and its contributions and limits. Indeed, multi-agent simulation increasingly attracts different communities because of the wealth and flexibility of solutions offered and also because of the degree of satisfaction of users already working with it.

Behavioral modeling was selected to represent the dynamic of a QoS network. We have described the characteristics of such a modeling and included some works by using behavioral modeling in different domains (collective robotics, life of an ant colony, etc.). A two-level behavioral model of a network node has been proposed. The first level is that of basic behaviors, whereas the second groups metabehaviors. The roles of each of these two levels as well as their characteristics were widely described. We have also presented the structure of the metabehavior rules responsible for the representation of node dynamics.

Some examples of basic behaviors and metabehaviors adapted to the networks with service differentiation as well as the results of multi-agent simulation based on the proposed model are given in Chapter 7.

6.7. Bibliography

[BAR 01] BARBER S., MARTIN C., "Dynamic Adaptive Autonomy in Multiagent Systems: Representation and Justification", *International Journal of Pattern Recognition and Artificial Intelligence*, vol. 15, no. 3, p. 405-433, 2001.

[BAY 00] BAYNAT B., *Théorie des files d'attente: des chaînes de Markov aux réseaux à forme produit*, Hermes, 2000.

[BAZ 03] BAZZAN A.L.C., KLÜGL F., "Route Decision Behaviour in a Commuting Scenario: Simple Heuristics Adaptation and Effect of Traffic Forecast", *Euroworkshop on Behavioural Responses to ITS*, Eindhoven, 2003.

[BEN 03] BENSAID L., Simulation multi-agent des comportements des consommateurs dans un contexte concurrentiel, Pierre and Marie Curie University, Thesis, June 13 2003.

[BLA 98] BLAKE S., BLACK D., CARLSON M., DAVIES E., WANG Z., WEISS W., "An Architecture for Differentiated Service", *RFC 2475*, December 1998.

[BOD 00] BODANESE E.L., CUTHBERT L.G., "A Multi-Agent Channel Allocation Scheme for Cellular Mobile Networks", *ICMAS'2000*, USA, IEEE Computer Society Press, p. 63-70, July 2000.

[BOW 01] BOWMAN H., BRYANS J.W., DERRICK J., "Analysis of a multimedia stream using stochastic process algebra", *The Computer Journal*, vol. 44(4), p. 230-245, April 2001.

[CON 98] CONTE R., GILBERT N., SICHMAN J.S., "MAS and Social Simulation: A Suitable Commitment", *MABS'98*, Paris, LNAI 1534, p. 1-9, July 1998.

[DOR 01] DORAN J., "Agent-Based Modelling of EcoSystems for Sustainable Resource Management", *3rd EASSS'01*, Prague, LNAI 2086, p. 383-403, July 2001.

[DRO 93] DROGOUL A., De la simulation multi-agent à la résolution collective de problèmes, Paris VI University, Thesis, November 23 1993.

[EDM 03] EDMONDS B., "Simulation and Complexity – how they can relate", in Feldmann V., Mühlfeld K. (ed.), *Virtual Worlds of Precision*, 2003.

[FAR 00] FARATIN P., JENNINGS N.R., BUCKLE P., SIERRA C., "Automated Negotiation for Provisionning Virtual Private Networks Using FIPA-Compliant Agents", *PAAM 2000*, Manchester, UK, p. 185-202, 2000.

[FER 03] FERBER J., GUTKNECHT O., MICHEL F., "From Agents to Organizations: An Organizational View of Multi-agent Systems", *AOSE*, p. 214-230, 2003.

[GIL 94] GILMORE S., HILLSTON J., "The PEPA Workbench: A Tool to Support a Process Algebra-based Approach to Performance Modelling", *7th International Conference on Modelling Techniques and Tools for Computer Performance Evaluation*, Vienna, LNCS 794, p. 353-368, May 1994.

[HER 01] HERZOG U., ROLIA J.A., "Performance validation tools for software/hardware systems", *Performance Evaluation*, vol. 45(2-3), p. 125-146, 2001.

[HIL 96] HILLSTON J., *A Compositional Approach to Performance Modelling*, Cambridge University Press, 1996.

[MAT 94] MATARIĆ M.J., Interaction and Intelligent Behavior, PhD Thesis, MIT EECS, May 1994.

[MER 03] MERGHEM L., MESKAOUI N., GAÏTI D., KABALAN K., "DiffServ network control using a behavioral multi-agent system", *NETCO proceedings*, Kluwer Academic Publishing, p. 51-63, 2003.

[MEU 01] MEURISSE T., VANBERGUE D., "Et maintenant à qui le tour ? Aperçu de problématiques de conception de simulations multi-agents", *ALCAA 2001*, Bayonne, September 2001.

[MGP 03] MERGHEM L., GAÏTI D., PUJOLLE G., "On Using Agents in End to End Adaptive Monitoring", *E2EMon Workshop/MMNS'2003*, Belfast, Northern Ireland, LNCS 2839, p. 422-435, September 2003.

[MOU 98] MOUKAS A., CHANDRINOS K., MAES P., "Trafficopter: A Distributed Collection System for Traffic Information", *CIA'98*, Paris, France, LNAI 1435 p. 34-43, July 1998.

[MUR 89] MURATA T., "Petri Nets: Properties, Analysis and Applications", *Proceedings of the IEEE*, vol. 77, no. 4, p. 541-580, April 1989.

[NIC 02] NICOLESCU M., MATARIĆ M.J., "A Hierarchical Architecture for Behavior-Based Robots", *1st ACM International Joint Conference on Autonomous Agents and Multi-Agent Systems (AAMAS'2002)*, Bologna, Italy, July 2002.

[PUJ 02] PUJOLLE G., "Les techniques "classiques" de modélisation et de simulation de réseaux et leur adaptation à la modélisation active", *Deliverable 11 from project MACSI*, February 2002.

[ROY 00] ROYCHOUDHURI R. *et al.*, "Topology discovery in ad hoc Wireless Networks Using Mobile Agents", *MATA'2000*, Paris, France, LNAI 1931, p. 1-15, September 2000.

[SER 00] SERVAT D., "Distribution du contrôle de l'action et de l'espace dans les simulations multi-agents de processus physiques", *JFIADSMA'2000*, Hermes, p. 343-356, October 2000.

[SIG 02] SIGEL E., DENBY B., LE HÉGARAT-MASCLE S., "Application of Ant Colony Optimization to Adaptive Routing in LEO Telecommunications Satellite Network", *Annals of Telecommunications*, 57, no. 5-6, p. 520-539, May-June 2002.

[VAN 02] VANBERGUE D., DROGOUL A., "Approche multi-agent pour la simulation urbaine", *Actes des Journées Cassini*, Brest, September 2002.

[WHI 99] WHITE T., PAGUREK B., "Distributed Fault Location in Networks using Learning Mobile Agents", *PRIMA'99*, Kyoto, Japan, LNAI 1733, p. 182-196, December 1999.

[WOO 02] WOOLDRIDGE M., *An Introduction to Multiagent Systems*, John Wiley and Sons, February 2002.

Chapter 7

Behavioral Modeling and Simulation: An Example in Telecommunications Networks

7.1. Introduction

In order to represent increasingly dynamic networks, a behavioral approach was chosen by several studies. The contribution of this approach in sectors other than networks and the inability of methodologies normally used such as queuing theory, process algebra, Petri nets, etc. have influenced this choice. However, and before declaring victory, we should prove the efficiency of this approach to study the performances of dynamic networks. In other words, it is necessary to verify the possibility of simulating a network based on the behavioral approach and also to make sure that the results are consistent. In this regard, we have completed the simulations of a DiffServ network with quality of service (QoS) in which each node contains metabehaviors, each dedicated to a given task and with the responsibility of choosing which basic behaviors to activate.

The goal of this chapter is to describe a behavioral approach application to model and to simulate a DiffServ network with QoS and to prove the ability of this approach to take into account the very dynamic nature of future networks. The rest of this chapter is explained as follows: we will begin with the description of basic behaviors adapted to networks. These basic behaviors are grouped according to their task. This means that we have basic behaviors for queue management, then for scheduling and then for routing. Section 7.3 describes queue management, scheduling and routing metabehaviors with examples of their metabehavior rules.

Chapter written by Leila MERGHEM-BOULAHIA.

The description of the simulation tool, agents and implemented objects as well as traffic generation parameters are given in section 7.4. A few results showing the performances obtained with basic behaviors and metabehaviors of queue management and scheduling are described in section 7.5 followed by a discussion in section 7.6. Section 7.7 concludes the chapter.

7.2. Basic behaviors adapted to networks

In this section we present the different basic behaviors that we have defined and implemented in order to respond to QoS requirements. To describe basic behaviors, we have grouped them according to the control task completed at node level. We start by presenting basic behaviors in queue management, followed by scheduling and finally routing. Other basic behaviors dedicated to admission control, failure detection, negotiation, etc. are discussed in [MER 03].

Before explaining the functions of proposed basic behaviors, it would be useful to remind our readers that they operate in a context of networks with differentiation of services. Three service classes, each requiring different QoS constraints, have been defined. Inspired by the DiffServ model [BLA 98], these classes are:

– *expedited forwarding* (EF or *premium*) class: packets in this class have the highest priority. They have total required service guarantee; they must not be lost or delayed;

– *assured forwarding* (AF or *olympic*) class: packets in this class must arrive at their destination (they must not be lost) but in a non-guaranteed time. This class may contain sub-classes, each one offering different delay and loss guarantees;

– *best effort* class: packets in this class have the least priority. In fact, these packets are only routed if the bandwidth is not occupied by higher priority classes, otherwise they are lost. The current Internet network only provides *best effort* services.

7.2.1. *Queue management basic behaviors*

Queue management is a router process which greatly influences network performance and especially affects the loss ratio of each traffic class and therefore the global loss ratio. Below is the description of four basic behaviors proposed for queue management: “careful”, “finder”, “suspicious” and “thrifty”.

7.2.1.1. *"Careful" basic behavior*

This basic behavior is inspired by the RED (*random early detection*) algorithm [FLO 93] and consists of placing packets in the queue according to their priority while observing the queue's occupancy level (Figure 7.1). When the queue load is at level 1 (the first MinTh threshold has not been reached yet), the node does not launch any special action except for placing packets in the queue according to their priority. As soon as the first threshold is reached (passage to second level), it starts to reject *best effort* packets coming from its sources (its own clients). From the second queue occupancy threshold (MaxTh) (third level), the node rejects all *best effort* packets, even those from other nodes, and sends a control message to its closest neighbors to inform them that its queue will soon be saturated. If a *premium* packet arrives, then the queue is full and the last *best effort* or even *olympic* packet in the queue is rejected in order to keep the most important packet. This is also valid if an *olympic* packet arrives when the queue contains *best effort* packets that it can reject to keep this packet.

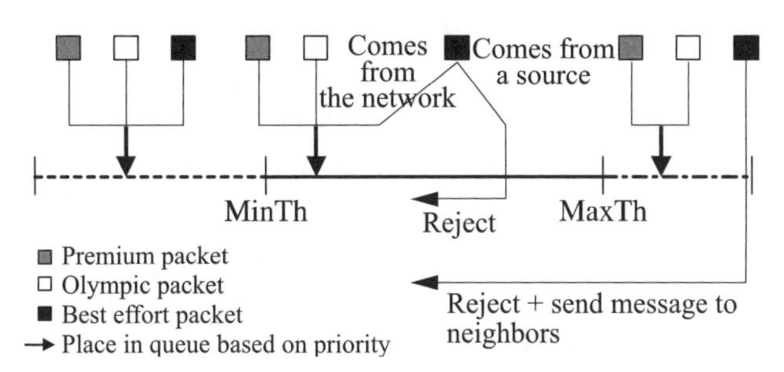

Figure 7.1. *Careful basic behavior*

7.2.1.2. *"Finder" basic behavior*

Finder basic behavior is a queue management basic behavior with the goal of guaranteeing QoS to flows crossing the node and to only penalize flows which are actually responsible for a possible congestion. It is inspired by the CHOKE algorithm [PAN 00], which is inspired by RED and which attempts to keep a stable average queue occupancy while trying to be fair and only target the flows that do not decrease their throughput during congestion.

Finder basic behavior operation is summarized in Figure 7.2. Its actions are always based on queue occupancy levels and it operates in the same way as careful basic behavior when queue occupancy is below the first threshold (placing packets in the queue according to their priority and not their time of arrival) or over the second threshold (rejecting all *best effort* arriving packets and sending a message to the surrounding nodes).

When queue occupancy is between both thresholds and a *best effort* packet coming from a source of the router arrives, another *best effort* packet is randomly drawn from the queue. If both packets belong to the same flow, they will both be rejected, otherwise the arriving *best effort* packet is accepted in the queue with a P probability that will depend on the current queue size. In the case where a *best effort* packet is rejected when the queue occupancy is between both thresholds, a message is sent to the packet's source. This message is significant in the sense that it informs the source that is it very probable that the packets it generates will be lost. The source's normal reaction would be to decrease its traffic until the queue load is below the first threshold again. At that moment, it will be able to start generating traffic at a normal rate.

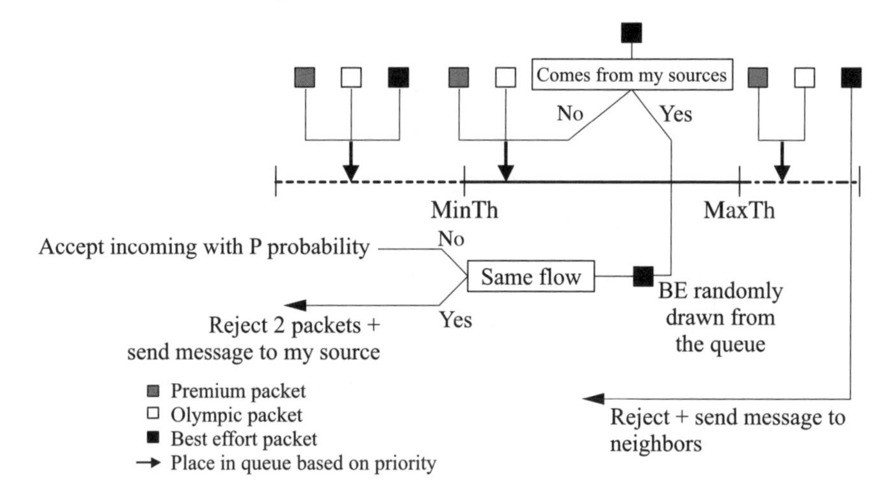

Figure 7.2. *Finder basic behavior*

7.2.1.3. *"Suspicious" basic behavior*

This basic behavior ensures the priority processing of flows coming from routers with the same basic behaviors. When congestion becomes imminent (MaxTh exceeded), it consists of rejecting or marking packets according to their origin. Rejection involves *best effort* packets, whereas marking involves *olympic* and *premium* packets whose last used router has not activated the suspicious basic behavior. Indeed, the node is supposed to know which basic behaviors have been activated among its neighbors. This can be achieved by messages sent during basic behavior changes so that neighbors get information reflecting reality.

The suspicious basic behavior operation is illustrated in Figure 7.3.

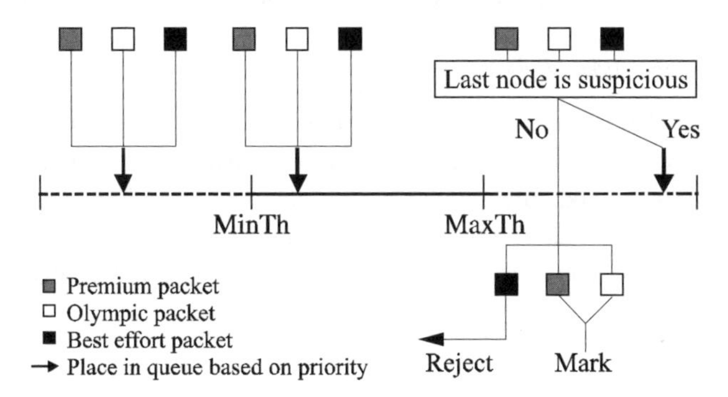

Figure 7.3. *Suspicious basic behavior*

Marking is actually only a deterioration of class. In other words, a *premium* class packet becomes *olympic* and an *olympic* class packet becomes *best effort*. The packet then does not get rejected but it does not get the best QoS either. On the other hand, it is presumed that marking is authorized by the QoS contract negotiated between the client generating traffic and the administrator.

7.2.1.4. *Economy basic behavior*

This basic behavior only keeps useful packets in the queue. It rejects packets belonging to a flow that does not support loss and that has already partly experienced reject. This comes from the principle that a flow which is partly lost will not give complete information. It then becomes unnecessary for the remaining packets of this flow to occupy network resources.

7.2.2. *Scheduling basic behaviors*

Scheduling is a mechanism that makes it possible to choose which queue will have its packets processed first when a router has several queues. The delay and jitter are two parameters which greatly rely on the scheduling policy [FER 01]. *Priority queuing* (PQ) [RON 97], *weighted fair queuing* (WFQ) [DEM 89], *round robin* (RR) [NAG 87], *customized queuing* (CQ) [VEG 01], etc. are scheduling algorithm examples discussed in other works and some of these examples have been a source of inspiration for our basic behaviors.

7.2.2.1. *Fair basic behavior*

This basic behavior provides a scheduling which processes all the queues fairly. Considering that the node has n queues, then queue X, X ϵ [1, n] will be served once over n cycles. This basic behavior is inspired by the basic *round robin* algorithm.

7.2.2.2. *Inequitable basic behavior*

This basic behavior will only serve a given queue if higher priority queues are empty. In this way, the lower priority queue may wait a long time for one of its packets to get processed. Inequitable basic behavior, even if it does not imply a notion of equality between the packets, is able to guarantee adequate performances in terms of delay, jitter and loss to priority packets. It is inspired by PQ [RON 97].

7.2.2.3. *Moderate basic behavior*

This basic behavior can be considered as a compromise between the two previous basic behaviors. It serves queues in a *round robin* way, but serves priority queues more often. This can happen in the following way: let us consider a router with n queues where the priority order is increasing (queue n has a higher priority than the n – 1 queue). The moderate basic behavior can have the following scenario: queue n is served n times before the n – 1 queue gets chosen by the basic behavior. This time (n – 1) packets from the current queue are processed. Only one packet from queue 1 can be processed once all the other queues with more priority are served. This basic behavior will help to avoid a phenomenon of famine, which might happen with inequitable basic behavior, while supporting queues with more priority by giving them more resources than the others.

7.2.3. *Routing basic behaviors*

Routing basic behaviors are responsible for forwarding packets to their destination. Several routing algorithms exist in other works and are implemented in routers such as RIP [HED 88], OSPF [MOY 91], etc. Some of them are dedicated to an internal or intradomain routing such as OSPF, whereas others are responsible for external or interdomain routing such as BGP [REK 95]. Two internal routing basic behaviors are given below.

7.2.3.1. *"Conservative" basic behavior*

This basic behavior ensures a static routing based on the choice of the shortest route whatever the state of traffic. It sends packets on the same route even if this route contains congested nodes which would increase the probability of packet loss. The advantage of this basic behavior resides in its simplicity.

7.2.3.2. *"Disloyal" basic behavior*

This basic behavior is also responsible for routing but completes a dynamic routing this time. In addition to holding the shortest route, it also holds secondary routes with more nodes that can be used in case of congestion. The objective is to keep the congestion from continuing by distributing the traffic load over sub-optimal routes. We should note that only packets with no delay constraint (*best effort* and *olympic*) will be sent through sub-optimal routes. *Premium* packets will continue to use the optimal route.

7.3. Metabehaviors

The node has a range of basic behaviors at its disposal to ensure a given control. Each of these basic behaviors can guarantee good performances under certain network conditions. Nevertheless, no basic behavior always gives the best performance whatever the network configuration, number of clients, nature of the traffic, etc. because these basic behaviors only process a reduced part of a given control task. They are basic and thus do not execute anything very complex such as the adaptation of their activation parameters and conditions. Their node management mechanisms must then be adapted to the new conditions. Because basic behaviors, by definition, do not have autoconfiguration possibilities, this task becomes the responsibility of metabehaviors. This section is devoted to the detailed description of each metabehavior responsible for node management, its role and the basic behaviors that it manages. Below we describe queue management, scheduling and routing metabehaviors.

7.3.1. *Queue management metabehavior*

This metabehavior must choose which queue management basic behaviors to activate and with what parameters. It uses metabehavior rules whose examples are given below. The first rule will referee between decisions from the different basic behaviors if several of them are activated (Chapter 6, section 6.5.5.2). In this example, the packet is rejected if at least one of the activated basic behaviors has made the decision. The other rules ensure the node's dynamism and adaptability. Rules 2 to 5 dynamically change careful, finder or suspicious basic behavior thresholds according to the obtained delays and losses, whereas rules 6 and 7 choose the relevant basic behaviors that must be activated (a simple process if all packets have the same traffic class (rule 6) or a differential process according to packet class (rule 7)).

1. If one of the basic behaviors has chosen the action reject then
reject the packet (restrictive action)
2. If MinTh is very large and the premium packet delay is large then
decrease MinTh by ***x****%*
3. If MinTh is very small and the queue occupancy from T (ms) is $< \alpha$ *% and the best effort packet loss* $> \beta$ *% then*
increase MinTh by ***x'****%*
4. If MaxTh is very large and the premium packet delay is large then
decrease MaxTh by ***y****%*
5. If MaxTh is very small and the best effort packet loss $> \alpha'$*% and the average queue occupancy on a T (ms) interval is* $< \beta'$*% then*
increase MaxTh by ***y'****%*
6. If (getClock() >10 p.m and getClock() < 8 a.m.) then
inhibit the current queue management basic behavior
activate FIFO
7. If the premium packet loss $\neq 0$ *or the olympic packet loss* $> \alpha''$*% then*
inhibit the current queue management basic behavior
activate careful (α, β)

7.3.2. *Scheduling metabehavior*

This metabehavior defines the method serving the node's different queues because of rules like the following:

1. If the loss of best efforts $> \alpha$ *% and the delay of premium* $< \beta$ *ms and the jitter of premium*$< \beta'$*ms then*
inhibit the current scheduling behavior
activate moderate
2. If the delay of premium $> \alpha$ *ms or the jitter of premium* $> \alpha'$*ms then*
inhibit the current scheduling behavior
activate inequitable
3. If the occupancy of the premium queue $< \beta'$*since T (ms) and the occupancy of the olympic queue* $< \beta''$*since T'(ms) then*
inhibit the current scheduling behavior
activate fair
4. If the occupancy of the premium or olympic or best effort queues = 0 since T''(ms) then
distribute the resources reserved for this queue over the other two

Rules 1 to 3 activate the most relevant scheduling basic behavior for current conditions and rule 4 will reattribute allocated resources to a given service class if no packet of this class is received in a certain time period.

7.3.3. *Routing metabehavior*

This metabehavior is responsible for the management of basic behaviors which make it possible to choose the route that the packet must use to arrive at destination. The following rules will react to congestion by activating the most appropriate basic behavior. In these two rules, message M1 is sent by a neighbor node when the 2nd occupancy threshold of its queue is exceeded (imminent congestion, hence the advantage of dynamic routing (disloyal)), whereas M2 is sent when the occupancy of its queue goes back down under the 1st threshold after a state of congestion:

1. If the message received is a type M1 (congested nodes in the neighborhood) then
inhibit current routing behavior
activate disloyal
2. If the message received is a type M2 (congestion remedied) then
inhibit current routing behavior
activate conservative

With this section, we finish the description of basic behaviors and metabehaviors adapted to a service differentiation network. We will now address the 2nd section of this chapter which is dedicated to the multi-agent simulation of this network.

7.4. Simulation components and parameters

In a network, the entities which must be represented are: nodes, links, sources, queues and packets. An entity can be represented, according to the role that it plays within the network, by an object (passive agent) or by an agent responsible for an intelligent control. Below, we describe different model agents and objects.

7.4.1. *Objects*

In our model, the simulation objects are packets, links and queues:

– links: two nodes A and B are connected by two unidirectional links (AB and BA). A link is explained by the following n-tuple: Link::<Identifier; Source_Node; Destination_Node>;

– packets: they move according to the node's decision to another node until they arrive at destination. A packet is explained by the following n-tuple: Packet:<Identifier; Identifier_flow; Source; Destination; Type; Size; State; Delay>. A packet (Identifier) is part of a flow of packets (Identifier_flow) of a given type (Type), which is transmitted from a given client (Source) and intended for another client (destination). It has a specific size (Size), may be in movement or stable (State) and takes time (Delay) to arrive at destination;

– queues: they contain packets that the node must process. A queue is represented by the following n-tuple: Queue::<Identifier; Capacity; Current_occupancy; index_premium; index_olympic>. The number of packets waiting to be processed is explained by the Current_occupancy parameter. The two last parameters index_premium and index_olympic indicate the indices of the last *premium* and *olympic* packets in the queue. These two parameters are mainly used by the basic behaviors which consider packet priority (careful, finder, etc.).

7.4.2. *Agents*

The agent classes that we have defined and implemented are explained below (Figure 7.4). We should note that other agent classes can be defined and added to the model. Certain classes have been implemented and were the subject of performance by simulation studies, whereas for other classes, only their tasks have been defined:

– client agent: represents a source. It is responsible for the generation of packet flows. Each client generates only one type of packet because of the Generator() method. The client not only sends packets, it also receives them. However, the only process that the received packets go through consists of putting them in the client's queue (in FIFO mode) while they are waiting to be serviced. For now, we do not execute a specific process on data received. The packet is erased as soon as it has been served;

– routing agent: processes packets in the queue. It sends packets to other nodes according to the routing table and their destination. It can have several basic behaviors such as conservative, disloyal, etc.;

– queue management agent: processes incoming packets and places them in the queue in compliance with their priority levels and with its current basic behaviors (careful, finder, suspicious, etc.);

– scheduling agent: decides which queue must be served first following the activated scheduling basic behavior (fair, inequitable, etc.);

– relational agent: is responsible for negotiation and cooperation operations with other nodes. It has not been implemented yet;

– guardian agent: makes sure that the number of simultaneous connections does not exceed a certain limit in order to guarantee the quality promised to clients. This agent has not been implemented yet;

– detector agent: is responsible for failure detection to bypass the faulty nodes and links. It has not been implemented yet;

– teacher agent: monitors several agents such as the routing agent, queue management agent, scheduling agent, etc. The teacher agent's metabehaviors choose the basic agent behaviors under its responsibility.

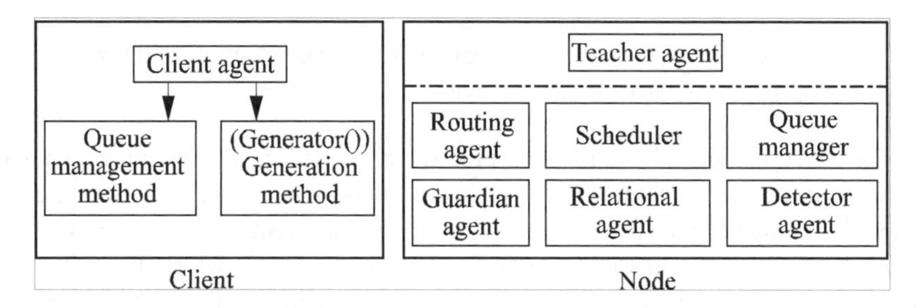

Figure 7.4. *Agents used in the simulation*

7.4.3. *Parameters*

The achievement of expected simulations implies the implementation of a series of basic behaviors and metabehaviors ensuring node resource control. It also leads to the choice of a multi-agent simulation environment, the choice of a network architecture, etc. Our choice has led us to oRis, i.e. a generic platform for the implementation of multi-agent systems, which is specifically dedicated to simulation [HAR 02]. However, we have had to implement all functionalities and components of a DiffServ network since oRis is not dedicated to telecommunications network simulation.

The topology that interests us is made up of six routers, each with four clients. Each router has a queue of 100 packets. A packet is generated by a client and is sent to another randomly chosen client. Packet generation is done in ON/OFF mode based on a Poisson process and the size of each generated flow follows an exponential rule. The generation parameters are different from one traffic class to another and can be found in [MER 03]. With the chosen parameters, the proportions of each traffic class are as follows: 25% *premium* packets, 50% *olympic* packets and 25% *best effort* packets. In reality, the administrator will not accept so many *premium* packets. By choosing these proportions, we want to prove that even under extreme conditions our basic behaviors and metabehaviors can ensure a good QoS for these packets.

7.5. A few results

The different simulations that we have completed can be classified in two categories:

– simulations where the nodes keep the same basic behaviors throughout the simulation,

– simulations where the nodes can change their basic behaviors and take up new, more appropriate ones through metabehaviors.

For each completed simulation, probes are installed on certain parameters (number of received packets, number of sent packets, delay on a given route, etc.) in order to compare the performances from the different basic behaviors and metabehaviors. Each simulation lasts for 10 minutes, after which the values of certain parameters are retrieved. QoS parameters, which will be the subject of optimization and will then become the comparison criteria between the different basic behaviors and metabehaviors, are the loss ratio of each service class as well as the delay and jitter from *premium* (and *olympic*) packets. For reasons of lack of space, we will just present the results obtained with two control tasks: queue management and scheduling. The results related to other control tasks such as routing and others showing the impact of different parameters such as queue size, proportion of each service class in global traffic, etc. are explained in [MER 03].

7.5.1. *Impact of queue management basic behaviors*

Several queue management basic behaviors have been defined and tested with activation parameters depending on simulation type and goal. The queue management basic behaviors tested are:

– careful basic behavior with the following values for its two thresholds: MinTh = 30, MaxTh = 50 for the first configuration, MinTh = 30, MaxTh = 70 for the second configuration and MinTh = 60, MaxTh = 90 for the third configuration. This enables us to see the impact of threshold choice over node performance;

– finder basic behavior with values of both thresholds equal to those of the careful basic behavior. In addition to seeing the impact of thresholds, this will enable us to compare these two basic behaviors (finder and careful);

– suspicious basic behavior with the values of both thresholds equal to those of the previously mentioned previous basic behaviors.

In Figure 7.5 it is possible to verify that the smaller the thresholds, the more significant is the loss of *best effort* packets. We can also notice that in all the configurations, finder shows less *best effort* packet loss than careful because it only rejects those that are really responsible for the congestion. Suspicious is the one that shows the least *best effort* packet loss because all the neighbors are suspicious. As expected, FIFO (*first in first out*) is the only one losing *premium* packets.

Loss of *best effort* packets represents the only criteria differentiating the tested basic behaviors. *Premium* packets are not lost (except in the case of FIFO) and *olympic* packets, even though they are in the majority in the network, have a low loss ratio and are almost equal for all configurations mentioned above. *Premium* and *olympic* packet loss is therefore neither related to the activated basic behavior (except with FIFO), nor to its activation parameters, but rather to the queue size and current occupancy.

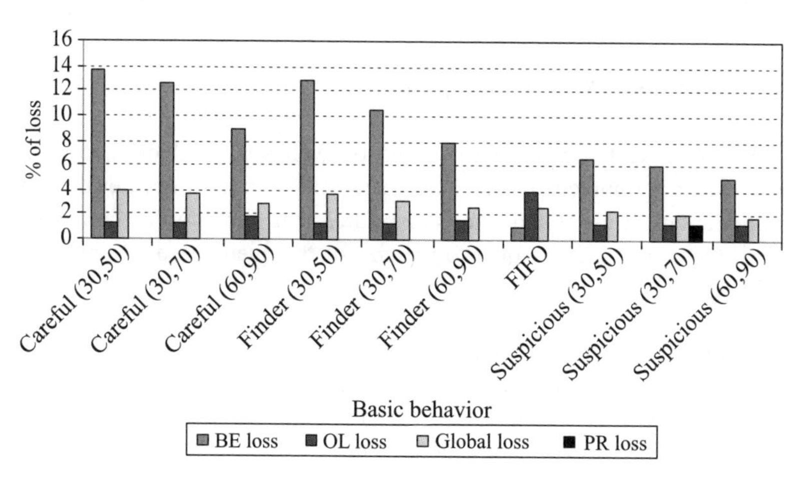

Figure 7.5. *Loss obtained with queue management basic behaviors*

Delay and jitter measures obtained with these series of simulations are summarized in Figure 7.6. It is possible to see that the delay and jitter of *olympic* packets increase with the increase of MinTh and MaxTh thresholds for the three basic behaviors tested (careful, finder and suspicious). We have also been able to verify that suspicious gives the highest delay and jitter measures because it is the one which has the least amount of *best effort* packet loss. The worst performances naturally come from the FIFO basic behavior.

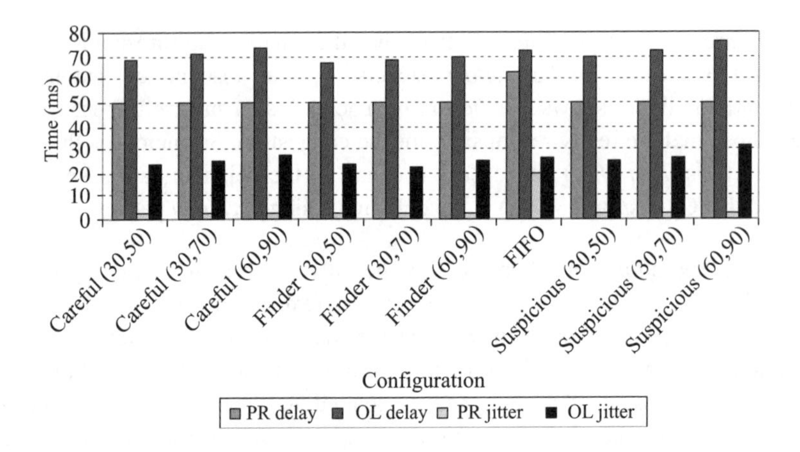

Figure 7.6. *Delay and jitter obtained with queue management basic behaviors*

7.5.2. *Impact of scheduling basic behaviors*

Three scheduling basic behaviors based on well known algorithms [RON 97, NAG 87] have been implemented. These basic behaviors have been evaluated with two configurations in which a queue is associated with each traffic class and where the sum of these file sizes is equal to the size of the queue used in simulations of section 7.5.1.

The first configuration applies the following sizes to the queues: 25 packets for *premium* class queue, 50 for *olympic* class queue and 25 for *best effort* class. For the second configuration, it uses the following sizes: 50 (*premium*), 40 (*olympic*) and 10 (*best effort*). The three queues are generated in FIFO mode, since the goal of these series of simulations is only to test scheduling basic behaviors.

Fair basic behavior is the behavior with the highest loss ratio of *olympic* packets, followed by moderate basic behavior. *Olympic* packet loss obtained with inequitable basic behavior is the lowest (Figure 7.7). The opposite is true for lost *best effort* packets, where the highest ratio is observed when inequitable basic behavior is activated. There is almost no loss of *premium* packets with inequitable and moderate basic behaviors and the loss is insignificant with fair basic behavior. This shows that, under current traffic conditions, a queue made up of 25 packets is not sufficient to process arriving *premium* packets.

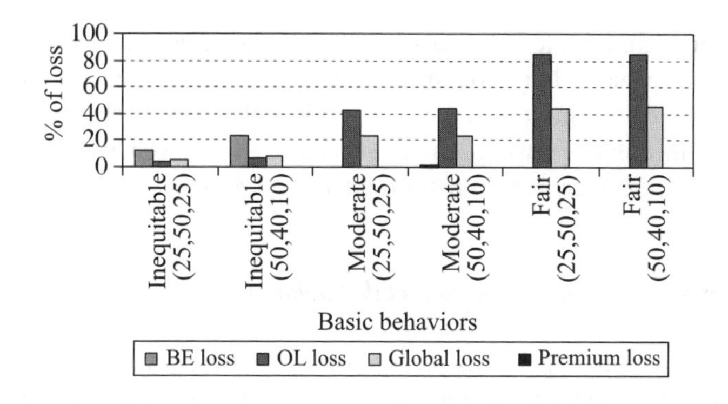

Figure 7.7. *Loss obtained with scheduling basic behaviors*

Inequitable basic behavior has better results when (25,50,25) configuration is used instead of (50,40,10). This result is obvious because in the second configuration, *olympic* and *best effort* queue sizes have been decreased in relation to the first one, thus causing more *olympic* and *best effort* packet loss and consequently more global loss. This assessment is the same for fair and moderate basic behaviors which experience an increased loss with the reduction of *olympic* and *best effort* queue sizes (configuration 2).

Figure 7.8 illustrates the delay and jitter measures of *premium* and *olympic* packets obtained with different scheduling basic behaviors. Inequitable basic behavior is the one with the best results since it first processes higher priority queues, thus enabling *premium* (and then *olympic*) packets to wait less time in the queues. The worst results are obviously obtained with fair basic behavior where packet priority is not a deciding factor for the choice of which queue to serve.

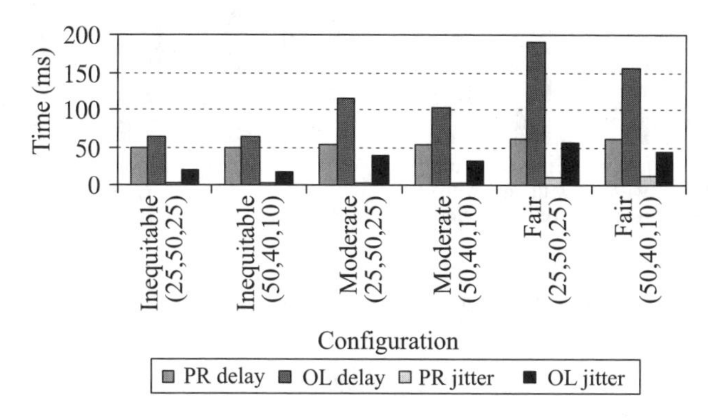

Figure 7.8. *Delay and jitter obtained with scheduling basic behaviors*

Fair and moderate basic behaviors result in higher delays and jitters when the size of *olympic* queue is larger (configuration 1); the packets are not lost but wait longer before being transmitted to their destination. *Premium* packet delay and jitter obtained with inequitable basic behavior remain the same whatever the size of the attributed queue (the delay is ≈ 49 ms and jitter is ≈ 1.4 ms).

7.5.3. *Impact of queue management metabehavior rules*

Through the simulations in this section we want to show the efficiency of queue management metabehavior rules (rules in section 7.3.1) and their ability to guarantee the best performance to clients.

The advantage of metabehaviors is mostly seen in a context where the nature of traffic is variable. In this context, the simulated scenario is: at simulation start-up, all the network nodes activate FIFO basic behavior or one of the other suspicious, finder, careful basic behaviors (with or without adding economy basic behavior) with random MinTh and MaxTh activation parameters. Traffic generation parameters change every 200,000 steps (200 s) by multiplying parameters of generation frequency (Poisson) or of flow length (exponentially) by 2.

Figure 7.9 shows that by using queue management metabehavior rules, network performance (loss, delay and jitter) is improved. These rules particularly influence MinTh and MaxTh parameters from the activated basic behavior of each node. In fact, the values of these thresholds experience increases or decreases in order to reach an appropriate compromise between packet loss and their delay and jitter.

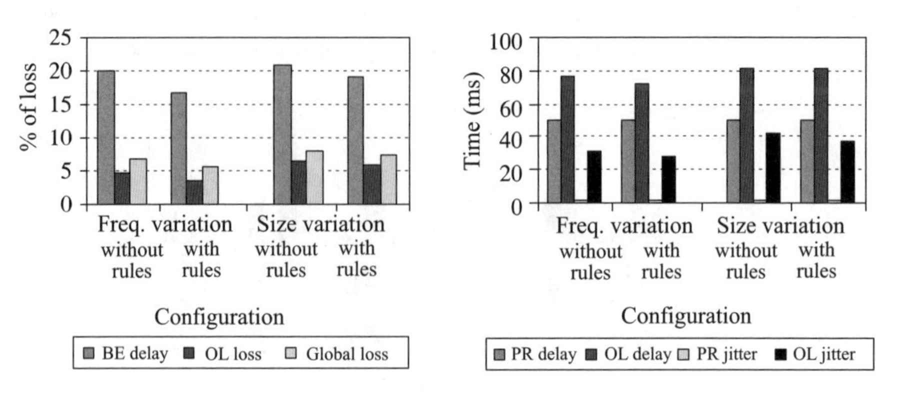

Figure 7.9. *Performance obtained with queue management metabehavior rules*

7.5.4. *Impact of scheduling metabehavior rules*

Inequitable basic behavior has shown good performances for *premium* and *olympic* packet delay and jitter. However, *olympic* packet loss remains relatively significant. This is explained by the static attribution of queues which does not consider the current traffic state. Queues of equal size (at least 25 packets) are attributed to *premium* packets (first configuration, section 7.5.2), whereas at certain times during simulation the network might not hold any. In order to fix this situation which causes some flow losses while other queues are practically empty, the scheduling metabehavior proposes rules that could optimize the use of the node's resources.

To clarify this metabehavior's contribution we have completed simulations representing highly irregular traffic situations in which the generation of one given type of traffic can become zero during a time interval and be very high during the rest of the time. Figure 7.10 reflects the scheduling metabehavior's resistance to these conditions by guaranteeing better performances (loss ratio, delay and jitter) than when a basic behavior is activated with the same parameters throughout the simulation.

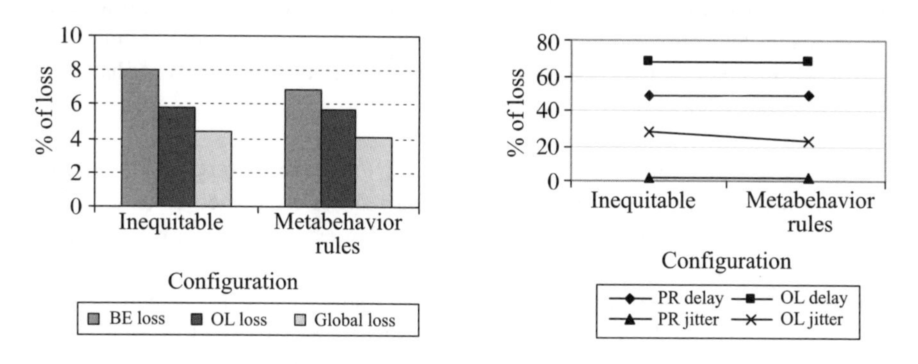

Figure 7.10. *Performance obtained with scheduling metabehavior rules*

7.6. Discussion

The goal of these simulations was to show the significance of using a behavioral multi-agent environment to explain the dynamic of new telecommunications networks. Simulation results, although expected, contribute to the verification of the accurate functioning of the completed simulation environment. In fact, each of the proposed basic behaviors was supposed to ensure a certain performance and we could see it in the results of the simulations accomplished.

We have in fact carried out several simulation scenarios where one section used parameters which do not reflect reality. Using such rigid and often "exaggerated" parameters makes it possible to test the capabilities of the proposed solution and see to what degree the system can be efficient.

The set of simulated basic behaviors have enabled us to understand the importance of choosing the correct activation conditions for the proposed basic behaviors as well as their activation parameters. We have in fact been able to study the performances obtained with several queue management and scheduling basic behaviors, and we have been able to verify their operation.

The performances obtained depend on activated basic behaviors and their activation parameters. We have been able to see the increase of losses with the decrease of thresholds (MinTh and MaxTh) on which careful, finder and suspicious basic behaviors base their processes.

Each basic behavior has its advantages and its drawbacks. Take, for example, careful and finder basic behaviors, which both guarantee good performances to *premium* and *olympic* flows. Finder basic behavior loses less *best effort* packets than careful basic behavior. On the other hand, it is more costly in processing time and presumes that the sources are cooperative and are equipped with some intelligence, even if rudimentary. An accurate compromise between proposed basic behaviors is vital and metabehaviors take care of this.

Tested metabehaviors, even by using few and simple rules, were able to optimize QoS parameters. The simulations that were carried out have tried to benefit from some of the results obtained with basic behaviors in order to arrive at better performances with the help of small modifications from certain parameters. Varying some MinTh and MaxTh thresholds, for example, leads to a better queue management, whereas by modifying the sizes of queues attributed to each service class, the scheduling metabehavior has also resulted in better performances.

The impact of using metabehavioral rules is positive since they use the services of basic behaviors optimally. On the other hand, there is a drawback in the use of such rules and that is cost. The choice and activation of metabehavior rules is a process where time increases with the number of available rules launched. In order to minimize the effects of this drawback, we have chosen to call on metabehavior rules every X simulation steps (1,000 for the simulations that we have done). With this interval there is enough time for the selected basic behaviors to make their modifications on the node's operation. A process like this avoids the appearance of performance peaks, which are signs of system instability.

In this chapter we have emphasized only the basic behaviors and metabehaviors responsible for two control tasks: queue management and scheduling. Nevertheless, in [MER 03] it is possible to examine the results obtained with routing basic behaviors and metabehaviors. It is also possible to see the impact of certain parameters such as the nature of the traffic received (its generation frequency, its size, the proportion of each service class), router characteristics (its speed, queue size) as well as the basic behaviors of the surrounding nodes, etc.

In addition, the behavioral model proposed is also used to refine the conditions of activation and settings for queue management and scheduling mechanisms in the context of a common study with Miguel De Castro from INT [DEC 04]. Completed tests and simulations in the scope of this study [MER 01, MER 02] have been integrated in the RNRT[1] MACSI[2] project. This has enabled us to make a comparison with other simulators such as ModLine from Simulog [SIM 03], which is a traditional network simulator. The comparison of results obtained with oRis and those obtained with Swarm [SWA 02], i.e. another multi-agent simulator, has also been done. The results obtained with different simulators were consistent. This proves that the model (objects, agents and their basic behaviors and metabehaviors) is correct.

7.7. Conclusion and perspectives

The proposed behavioral modeling makes it possible to represent as closely as possible the main features of new networks such as the adaptation of services provided to traffic state, the dynamic and adaptive aspect of the methods used to guarantee QoS required by users, heterogenous traffic transmitted over the network, etc. This is obtained through the adaptive and dynamic selection of basic behaviors to activate by taking into consideration the current network state. The simulations done have shown the advantage of our approach to find the correct basic behavior combinations to activate, as well as the effectiveness of the chosen model to represent the dynamic of new networks. The model that we have proposed effectively responds to environment variations and perfectly adapts to new traffic conditions.

The perspectives of this chapter mainly involve defining and testing basic behaviors and metabehaviors which are responsible for admission control,

---

1 Réseau national de recherche en télécommunications (national network of research in telecommunications).

2 Modélisation active et simulation des réseaux de télécommunications (active modeling and simulation of telecommunications networks).

negotiation, etc. within a DiffServ network as well as in other types of networks (MPLS [ROS 01], etc.). The learning constitutes an interesting channel to explore and which is liable to improve node performances by anticipating the actions of clients and/or surrounding nodes, etc. We also believe that the cooperative aspect of different routers (from one domain or belonging to two adjacent domains) can be increased by giving them cooperation algorithms [AKN 01].

7.8. Bibliography

[AKN 01] AKNINE S., PINSON S., SHAKUN M.F., "Négociation multi-agent: analyse, modèle et expérimentations", *Revue d'intelligence artificielle, RIA*, Hermes, vol. 15, no. 2, p. 173-218, 2001.

[BLA 98] BLAKE S., BLACK D., CARLSON M., DAVIES E., WANG Z., WEISS W., "An Architecture for Differentiated Service", *RFC 2475*, December 1998.

[DEC 04] DE CASTRO M.F., MERGHEM L., GAÏTI D., M'HAMED A., "The Basis for an Adaptive IP QoS Management", *IEICE Journal*, Special Issue on Internet Technology, series IV, 2004.

[DEM 89] DEMERS A., KESHAV S., SHENKER S., "Analysis and Simulation of a Fair Queuing Algorithm", *ACM SIGCOMM Symposium on Communications Architectures and Protocols*, Texas, USA, p. 1-12, September 1989.

[FER 01] FERRARI T., PAU G., RAFFAELLI C., "Measurement Based Analysis of Delay in Priority Queuing", *IEEE Global Telecommunication Conference Globecom'2001*, San Antonio, USA, November 2001.

[FLO 93] FLOYD S., JACOBSON V., "Random Early Detection Gateways for Congestion Avoidance", *IEEE/ACM Transactions on Networking*, vol. 1, no. 4, p. 397-413, August 1993.

[HAR 02] HARROUET F., TISSEAU J., REIGNIER P., CHEVAILLIER P., "oRis: un environnement de simulation interactive multi-agents", *Technique et science informatiques*, vol. 21, no. 4, p. 499-524, 2002.

[HED 88] HEDRICK C., "Routing Information Protocol", *RFC 1058*, June 1988.

[MER 01] MERGHEM L., LECARPENTIER H., GAÏTI D., "Définition des comportements avancés", *Deliverable 6 of project MACSI*, December 2001.

[MER 02] MERGHEM L., GAÏTI D., "Behavioural Multi-agent Simulation of an Active Telecommunication Network", *Starting Artificial Intelligence Researchers Symposium (Stairs'2002, in conjunction with the 15th European Conference on Artificial Intelligence)*, Lyon, IOS Press, p. 217-226, July 2002.

[MER 03] MERGHEM L., Une approche comportementale pour la modélisation et la simulation des réseaux de télécommunications, PhD Thesis, Paris 6 University, 27 November 2003.

[MOY 91] MOY J., "OSPF Version 2", *RFC 1247*, July 1991.

[NAG 87] NAGLE J., "On packet switches with infinite storage", *IEEE Transactions on Communications*, vol. 1, April 1987.

[PAN 00] PAN R., PRABHAKAR B., "A Stateless active Queue Management Scheme for Approximating Fair Bandwidth Allocation", *IEEE INFOCOM*, March 2000.

[REK 95] REKHTER Y., LI T., "A Border Gateway Protocol", *RFC 1771*, March 1995.

[RON 97] RONNGREN R., AYANI R., "A comparative study of parallel and sequential priority queue algorithms", *ACM Transactions on Modelling and Computer Simulation*, vol. 7, no. 2, p. 157-209, 1997.

[ROS 01] ROSEN E., VISWANATHAN A., CALLON R., "Multiprotocol Label Switching Architecture", *RFC 3031*, January 2001.

[SIM 03] http://www.simulog.fr.

[SWA 02] http://www.swarm.org.

[VEG 01] VEGESNA S., *IP quality of service*, Cisco Press, 2001.

Chapter 8

Multi-agent System in a DiffServ Network: Behavioral Models and Platform

8.1. Introduction

Traffic in telecommunications networks and the nature of information are increasing exponentially, which requires strict constraints in terms of throughput, delay and quality of service (QoS). Most traditional networks are not adequate to respond to these new requirements as their architectures remain relatively passive, inflexible and are not very progressive. Integrated solutions such as IntServ [BRA 94], RSVP [WRO 97], DiffServ [BLA 98] and RTP/RTCP have been proposed to respond to some extent to user needs. However, they may become obsolete in the near future.

This chapter studies the possibility and the advantage of integrating new approaches based on the notion of agents and multi-agent systems in telecommunications networks – specifically DiffServ networks [BLA 98]. Multi-agent systems are based on the notion of agents. An agent is an intelligent entity, having its own skills, which is able to interact with other agents in its domain for a specific goal. Multi-agent systems are made up of a distributed organization of agents, which provides a flexibility enabling a better reactivity and adaptability to changing and distributed situations.

Section 8.2 presents and reviews some proposed network solutions. Section 8.3 addresses the notion of agents and multi-agent systems and discusses the advantage

---

Chapter written by Nada MESKAOUI.

of their integration within networks. Our proposal of integrating agents in a DiffServ network is explained in section 8.4. Section 8.5 describes the simulation platform used and illustrates simulation results. Finally, our conclusion is the subject of section 8.6.

8.2. Quality of service – existing solutions and their problems

The Internet is a network interconnecting several heterogenous networks. This heterogenity comes from the fact that each network defines its own communication protocol between its nodes. The Internet protocol (IP) ensures the masking of this heterogenity and enables exchanges between these different networks. The service offered by IP is based on the transmission mechanism with no connection or error recovery. Packets can arrive in any order and with variable time delays and because of this the IP service is qualified as *best effort*.

With the growth of multimedia applications, traffic in telecommunications networks and the information format increase exponentially and bring increasingly strict constraints in terms of throughput, delay and loss rate and jitter. The basic IP network is not adequate to respond to these requirements because it is relatively passive and inflexible. In order to satisfy users, roughly integrated solutions occupy a large part of the current research in this field. These research projects are based on three approaches which help to improve quality of service:

– adaptation of applications behavior;

– network resource reservation;

– adaptation of network behavior.

Several tools and techniques have been designed in order to implement these different approaches, such as RTP/RTCP, IntServ/RSVP and DiffServ.

8.2.1. *RTP/RTCP*

The *real-time protocol* (RTP) was proposed to respond to real-time application requirements, such as digital voice and videoconferencing, which require a quality of service that the traditional Internet cannot provide. This protocol takes care of real-time management as well as management of multicasting sessions.

The problem with this mechanism resides in the fact that it is not possible to act on network nodes and to modify their processes, so the application will modify its behavior according to the network state. The network state's measures will be sent back to the application by RTP/RTCP: RTP packets transport software information, whereas RTCP packets transport monitoring information.

8.2.2. *IntServ/RSVP*

The IntServ [BRA 94] architecture proposes an extension of the standard Internet to integrate new services without fundamentally modifying its basic architecture. This type of network implements the notion of quality of service (QoS) in each of its nodes to ensure that each flow transmits according to the required service model. This requires a reservation of resources for each flow transmitting in the node. The reservation protocol used by IntServ is RSVP.

RSVP (*resource reservation protocol*) [WRO 97] is a signaling protocol with the responsibility of alerting the intermediate nodes to flow arrivals corresponding to specified QoS. The RSVP protocol does resource reservation in the nodes from the receiver or receivers (in the case of multicasting) back to the sender. In this case, when a new point is added to the multicast, it can perform the reservation in a more simple way than it could be done by the transmitter.

However, the RSVP protocol has a scaling problem. We must have as much status information as we have applications communicating through the node. This is a drawback for RSVP and IntServ which uses it. For this reason, we consider that this protocol is mainly adapted to a network where the number of computers is limited, such as a company network. It is in fact possible to use IntServ/RSVP for long distances if we know that the packets will pass through networks which guarantee QoS such as ATM. However, over IP it is not so easy to deploy IntServ/RSVP in a large-scale network and to resolve this problem, the DiffServ architecture has been proposed.

8.2.3. *DiffServ*

Diffserv [BLA 98] is an architecture meant to ensure a certain large scale network QoS. The approach adopted to implement the differentiated services is the redefinition of the TOS byte of IPv4 or the *traffic class* byte of IPv6. These fields are then used to mark arriving packets and to give each of them a specific service level, which results in a sufficient allocation of memory resource, bandwidth, etc. at node level in order to ensure the requested service. When entering a network, the traffic is labeled and sorted, and the internal nodes process the packets by their quality level defined by the *per hop behavior* (PHB). Three PHBs are proposed by DiffServ: *expedited forwarding* (EF) [JAC 99] which guarantees a minimum delay, loss and jitter; *assured forwarding* (AF) [HEI 99] which ensures different levels of QoS guarantees in terms of memory and bandwidth resources; and finally *best effort* (BE) which represents the standard service offered by the Internet with no guarantee.

DiffServ proposes an extension of the standard structure of a node with components which make it possible to ensure DiffServ functionalities. These components are: *classifier*, *meter*, *marker* and *shaper/dropper*. The *classifier* filters the packets received into different service classes. The *meter* measures the time properties of a flow to determine if it complies with the requested service. The *marker* marks the packets with a *DiffServ code point* (DSCP) which determines their PHB. Finally, the *shaper/dropper* makes it possible to delay or lose some packets belonging to a given flow.

The advantage of DiffServ comes from the aggregation of flows requiring the same QoS. This involves reservations and processing by service type instead of by flow, thus facilitating remote DiffServ deployment. However, DiffServ has two problems: inaccuracy to adapt application requirements to network capacity and the use of standard routing algorithms which choose the shortest route to reach a destination with no consideration for network load or the possibility of finding other alternatives to reach the destination without risk of congestion [HUS 00].

8.3. Agents, multi-agent systems and architectures

Advanced techniques based on the notion of agents and multi-agent systems (MAS) have long proven their effectiveness in several domains. During the last few years, these techniques, especially mobile agents, have been proposed in the telecommunications network field to resolve complex problems. The benefits gained from the implementation of agents and MASs in these networks have reinforced the research efforts in this domain.

8.3.1. *Agents*

The notion of agent is of drastic importance in artificial intelligence. This importance has been growing since the 1980s. Several designs, interpretations and understandings are associated with the term agent. We can observe two different agent concepts: one given by AI experts and one found in software marketing, generally called network agent.

For [DEM 91], in the AI world the term agent means intelligent entity acting in a rational and intentional way according to its goals and the current state of its knowledge. [SHO 93] defines an agent as a high level entity which can be controlled and which acts in a continuous and autonomous way in an environment where processes take place and where other agents exist. [MIN 94] generally considers the agent as a person working and managing an agency, a substance provoking a chemical reaction, an organism provoking an illness, etc.

[FER 97] defines an agent as being a physical or virtual entity:

– able to act in an environment;

– able to communicate directly with other agents;

– driven by a set of directions;

– with its own resources;

– able to perceive (in a limited way) its environment;

– with only a partial view of its environment;

– with skills and offering services;

– able to reproduce in the future;

– whose behavior tends to satisfy its objectives by taking into account the resources and skills which it has and according to its perception, its representations and the communications it receives.

According to [GUE 96], the overview of these definitions confirms that the question "what is an agent?" is as awkward as "what is intelligence?".

It is preferable to not give a definition of an agent at all so that we do not limit the scope of this field. On the other hand, concerning the notion of network agents, [BOU 98] proposes a classification in three categories, mobile agents, information agents and interface agents and we propose a fourth category, control agents:

– mobile agents are autonomous, proactive and reactive entities taking the place of users to interact with servers. Several projects have been launched around this concept in order to benefit from the autonomy and mobility of these agents to support, for example, user mobility [THA 99] and telecommunications network control [APP 94];

– information agents take on the role of management, operation and retrieval of distributed information. An example of information agents is the Softbot [ETZ 94] agent which accepts high level requests from a user and is able to research and gather knowledge in order to determine how to satisfy these requests;

– interface agents act as autonomous personal assistants cooperating with users to complete specific tasks. It monitors the user's actions, offers help in using applications and suggests the best ways to complete a task;

– control agents, in our case, take care of controlling telecommunications networks to improve their performance. These agents are implemented in network nodes to control their behavior and make appropriate decisions according to the network state. These agents offer autonomy to nodes mainly in the resolution of failures such as congestion problems.

In order to display intelligent behavior, an agent needs knowledge. This knowledge is the result of its interaction with other agents and with the outside world. The agent also possesses beliefs and commitments towards itself and its environment. This belief defines information that is part of an agent's world, whether this information is true or false. Beliefs are the structure, state and evolution law of the environment and the state and behavior of agents.

8.3.2. *MAS*

An MAS can be seen as a group of agents interacting in an environment for the purpose of reaching a global objective. This interaction is performed by a communication or even a cooperation and coordination between agents.

Communication in an MAS is not reduced to a simple process of message transmission/reception, but it is also an act of communication. According to [FER 97], communicating is a particular form of action which, instead of transforming the environment, tends to modify the recipient's mental state. [BUR 92] also considers that messages between agents must be structured in such a way that the transmitter's intention should instantly be recognized from the message itself.

For an efficient communication operation, cooperation between agents is required. This cooperation will ensure a global consistency from the local decisions of agents. Cooperation as defined by [GAL 88] introduces three vital conditions for cooperation: recognizing the goal of another agent, be involved in its resolution by adopting it as its own goal in a deliberate way and consider it as a common goal. Knowing that this goal is also the goal of another agent is a vital element of cooperation. In this way, cooperation is more than an accidental coordination; it represents a support for the communication. A consequence of these conditions is that there is a commitment to achieve the goal.

According to [DUR 89], there are four generic cooperation goals: 1) increasing task resolution speed by their parallelization, 2) increasing all or the range of tasks achievable by resource sharing, 3) increasing the probability of completing tasks (reliability) by task duplication and if possible by the use of different methods, 4) decreasing interference between tasks by avoiding damaging interactions.

The multi-agent approach is a general method for the modelization and resolution of complex problems. MAS, due to their distributed nature and the adaptive aspect of their integrated agents, constitute architectures which are particularly able to consider the growth and flexibility required for network operation.

Among the applications implementing MAS, we can mention, for example, road traffic control with driver assistance for choosing the best route [MOU 98], the study of user behavior in a competitive market [BAZ 99] and the optimization of the routing process in a satellite constellation [SIG 02].

8.4. Towards intelligent and cooperative telecommunications networks

Different network solutions have been proposed in the last few years to improve telecommunications network performance and to provide users with a better QoS. These techniques, each with its own objectives, have been able to introduce improvements in network operations despite some limitations described in section 8.2. To overcome these limitations, the new generation of networks must implement new techniques able to make the best use of existing solution advantages.

We believe that future networks must be autonomous, implementing mechanisms which are able to ensure scaling, adaptability and network survival. These characteristics can be carried out by mechanisms inspired by the artificial intelligence field which implies that advanced techniques integrating agents and MAS in telecommunications networks can be considered as promising approaches for handling network problems.

Below, we propose a generic solution implementing an agent in each network node. This defines an MAS in which agents communicate and cooperate for the purpose of reaching a common objective. Each of these agents is also able to communicate with the different components of its node to retrieve information, modify parameters and act on the router's global behavior according to the state of the network.

8.4.1. *Node structure*

Our approach is to integrate notions of intelligence and collaboration in telecommunications networks. These notions are implemented by an agent with a behavior designed to reach a specific objective. The structure of a standard node implementing this agent is presented in Figure 8.1.

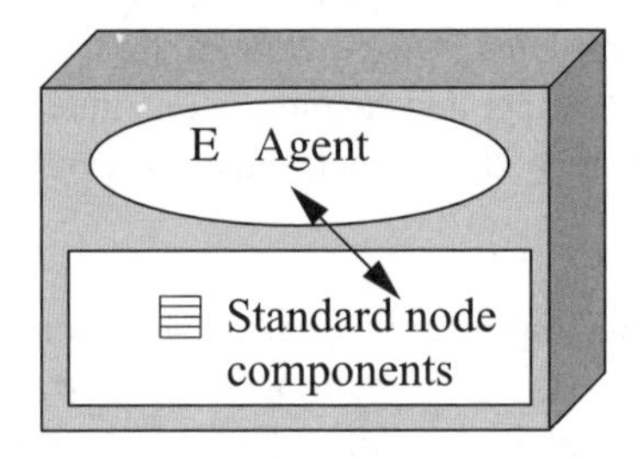

Figure 8.1. *Structure of a node implementing an agent*

The agent integrated in the standard structure of a node has intelligent capabilities enabling it to react to changes in its environment and to act on node components to modify the global network behavior according to these changes.

To validate our proposal we have chosen to implement the agents in a DiffServ network with the objective of preventing and stop congestion [MES 03a, MES 03b]. We have chosen the DiffServ network to benefit from its QoS and to try to define an agent behavior which is able to improve the performance of a network with QoS instead of integrating differentiation between different types of services within the agent's behavior, thus making it more complex.

Figure 8.2 shows the structure of a node on two levels: the first implements the different DiffServ components – *classifier*, *meter*, *marker* and *shaper/dropper* – and the routing algorithm. The second implements agent components.

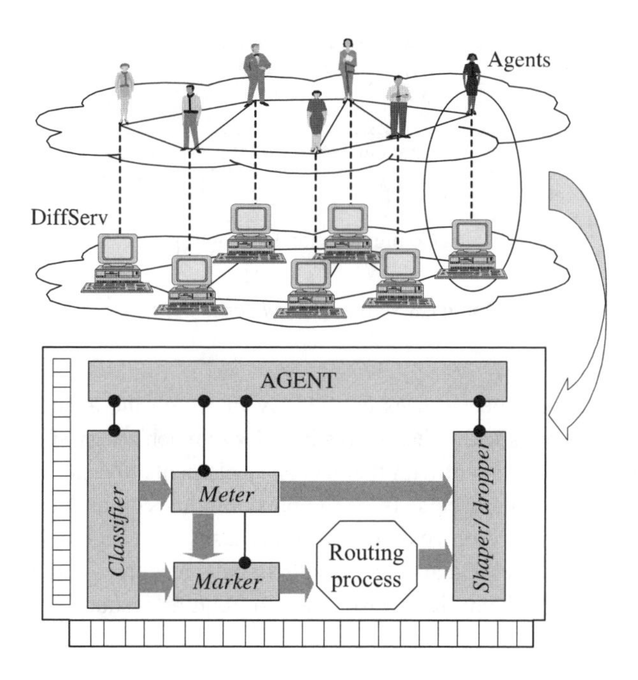

Figure 8.2. *Structure of a DiffServ node implementing an agent*

The different agent module components are described in the following section.

8.4.2. *Agent components*

The proposed agent model is broken down into several components or aspects – described in Figure 8.3 – which are intelligence, cooperation, knowledge and interagent communication language.

The knowledge base represents the knowledge acquired by the agent during its environment perception and its communication with other agents within its system. This knowledge takes part in the determination of agent behavior.

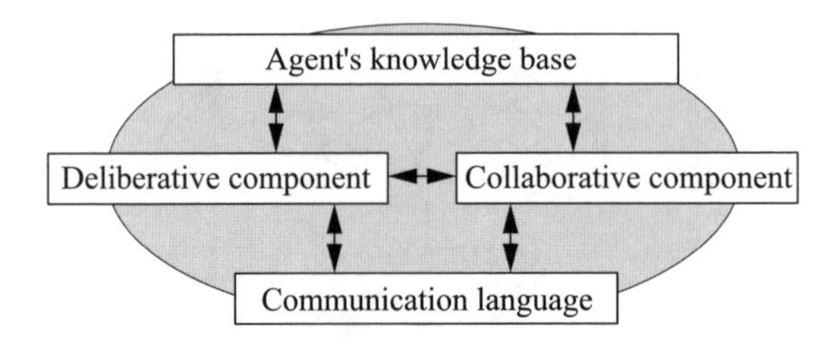

Figure 8.3. *Agent model components*

The collaborative aspect processes the way in which the agent communicates, interacts and cooperates with other agents in order to reach a very specific goal. This component provides the agent with the necessary elements to establish and maintain conversations with other agents. In the context of this communication, the agent demands and offers resources representing tasks to accomplish or simply information. We consider that an agent is able to have resources and to communicate its capacity to offer them; it is also able to require resources and search for providers.

The intelligent or deliberative component implements the agent's behavior which reflects its intelligence level. We have chosen to adopt a reactive behavior for our agent model. The reactive agent reacts to changes in its environment or to receiving a message from another agent. Other types of agents representing a higher intelligence level could have been implemented, but we have chosen to test the advantage of integrating agents in a telecommunications networks while using simple behaviors. Other more complex behaviors, showing higher intelligence, can be introduced in the future.

The interagent cooperation requires the specification of a communication language comprehensible by all MAS agents. This aspect is specified in the "communication language" component which adopts the theory of speech acts. All these components interact to determine the agent's global behavior.

8.4.3. *Agent behavioral model*

Different behavioral models can be defined to specify the global behavior of an agent integrated in a telecommunications network. Below we propose two reactive behavioral models with the objective of prevention and resolution of congestion problems in a DiffServ network. These two models have a common goal but their level of complexity/intelligence is different.

8.4.3.1. *Model 1*

The first model proposes a very simple behavior which consists of controlling the different output queues installed in a DiffServ node. If the fillrate of one of these queues reaches a certain level "L" – specified by the agent according to the state of the network – the agent reacts to prevent congestion. This reaction consists of sending messages to all its neighbors to demand a decrease in the traffic rate, which is interpreted by the receiver as a modification in the routing table to forward some of their flows towards other output links – if possible.

This model specifies two types of behaviors: "careful" and "cooperative". The "careful" behavior [MER 03], which is described in Figure 8.4, represents the agent's cautiousness to prevent congestion and then to react before the node becomes congested. This behavior is explained by two production rules, "PRUD1" and "PRUD2", transmitting between two different states "P0" and "P1".

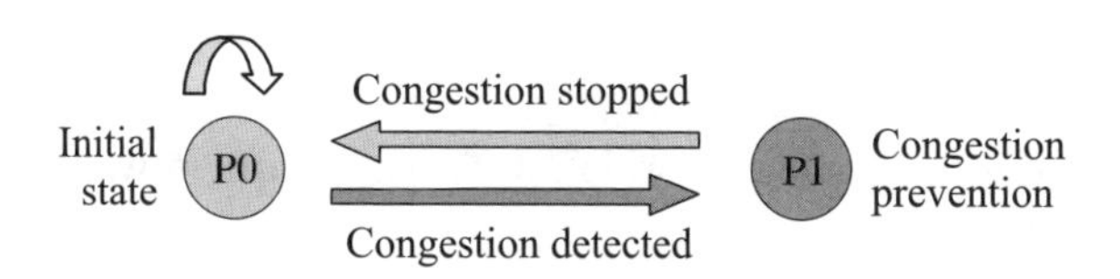

Figure 8.4. *Careful behavior for model 1*

"P0" represents the agent's state where no risk of congestion is reported and "P1" represents the state of an agent trying to process or prevent congestion. Both rules are described in the following way.

PRUD1: IF (P0) AND (a risk of congestion is reported) THEN:

– transition to (P1);

– send messages to neighbors to inform them of the risk and request help.

PRUD2: IF (P1) AND (there is no longer a risk of congestion) THEN:

– transition to (P0);

– send messages to neighbors to inform them of the removal of risk.

The "cooperative" behavior, described in Figure 8.5, represents the agent's intention to cooperate with its neighbor to prevent a possible congestion. This behavior is described by two production rules, "COOP1" and "COOP2", transmitting between two different states: "C0" and "C1".

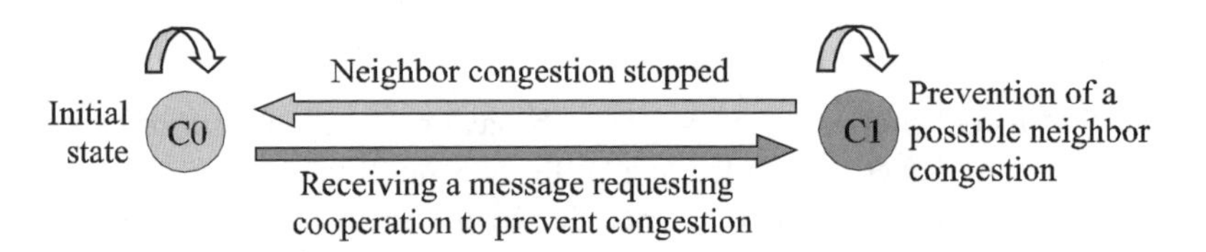

Figure 8.5. *Cooperative behavior for model 1*

"C0" represents the agent state when no risk of congestion from a neighbor is reported and "C1" represents the state of an agent trying to cooperate with a neighbor who has detected a risk of congestion. Both rules are described in the following way.

COOP1: IF (C0) AND (reception of a message reporting a neighbor congestion) THEN:

– transition to (C1);

– modification of the routing algorithm to forward traffic towards the available connections.

COOP2: IF (C1) AND (reception of a message reporting that there is no longer a risk of congestion) THEN:

– transition to (C0);

– initialization of the routing table to default connections.

This model shows good results – illustrated in section 8.5.3.3 – but there remains a risk that congestion will return, after neighbors report a resolution of the risk, if all sources continue to send the same rate of traffic. This phenomenon comes from the fact that all neighbors redirect their traffic to the default route simultaneously.

8.4.3.2. *Model 2*

The second model proposes a more intelligent and complex behavior than the first behavioral model. This behavior consists of controlling the load of each output connection instead of the different DiffServ queues installed for each output connection – as in the first model – which gives the agent a global view of the load so that it can react in a quicker way. For each connection, two congestion alert levels "L1" and "L2" are defined by the agent according to the network status. If the load rate of a connection reaches the first level, the agent reacts by launching a search request for neighbors able to establish routing changes. When the agent receives the request – if it has rerouting possibilities – it responds by indicating a traffic type "EF or AF" for which it is able to modify its transmission and the cost of this transaction. When the initiating agent receives the responses, it updates its database and registers the received information. When the second level "L2" is exceeded, the agent that acquired the information decides, according to the neighbors' responses, which agents must modify their routing table. This decision is sent to the agents involved for immediate execution. When the risk of congestion is resolved, the agent notifies these agents in stages about the new state of its node so that they can again modify their routing tables by choosing the cheapest routes. The expression "in stages" means that the agent first notifies the neighbor executing rerouting at the highest cost and then the others until level L2 is reached again. This helps to prevent a second risk of congestion.

Modifying the routing decision is done as in the first model by modifying the costs of the links involved.

This model defines two types of behaviors: "careful" and "cooperative" [MER 02]. The "careful" behavior described in Figure 8.6 represents the cautiousness of the agent to prevent congestion and then to react before the node becomes congested. This behavior is explained by four production rules, "PRUD1" "PRUD2" "PRUD3" and "PRUD4", transmitting between three different states: "P0", "P1" and "P2".

"P0" represents the state of the agent where no congestion risk is reported. "P1" represents the state of an agent that has detected an overrun of "L1" and attempts to retrieve the information concerning the ability of the neighbors to establish rerouting for some types of DiffServ traffic.

"P2" represents a state of urgency in processing a risk of congestion. If information retrieval established in "P1" is finished before reaching "P2", the agent uses this information and the state of the link having the risk of congestion to choose the nodes which must execute rerouting and asks them to do it. In the case where level "L2" is reached before the end of information retrieval, the action executed by the agent in "P1" is interrupted for the transition to "P2". In this case, the agent sends help requests to all its neighbors as in the first model.

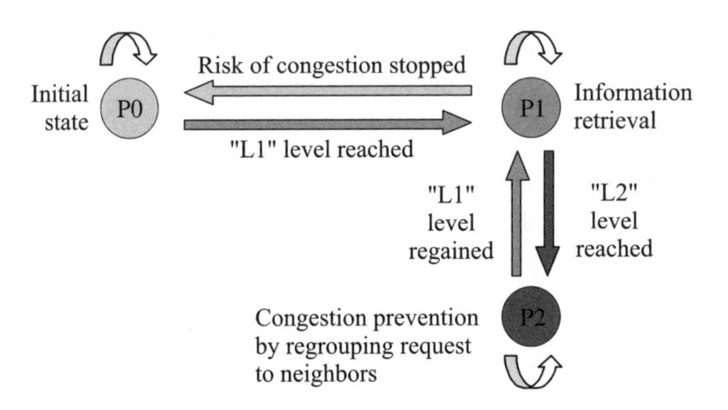

Figure 8.6. *Careful behavior of model 2*

The four rules are described below.

PRUD1: IF (P0) AND (level "L1" is exceeded) THEN:

– transition to (P1);

– send a request to neighbors for finding out the neighboring nodes capacity to establish rerouting for a certain type of traffic.

PRUD2: IF (P1) AND (level "L2" is exceeded) THEN:

– transition to (P2);

– send messages to specific neighbors (if information retrieval is finished) or to all the neighbors (if P1 action is interrupted) to ask them to execute rerouting.

PRUD3: IF (P2) AND (connection load is between L1 and L2) THEN:

– transition to (P1);

– send a new request to neighbors to update capacity information from neighboring nodes for rerouting a specific type of traffic.

PRUD4: IF (P1) AND (connection load is less than "L1") THEN:

– transition to (P0);

– send messages for agents having established rerouting to report the end of the congestion risk.

The "cooperative" behavior described in Figure 8.7 represents the agent's intention to cooperate with its neighbors to prevent a possible congestion. This behavior is described by five production rules, "COOP1" to "COOP5", transmitting between three different states: "C0" "C1" and "C2".

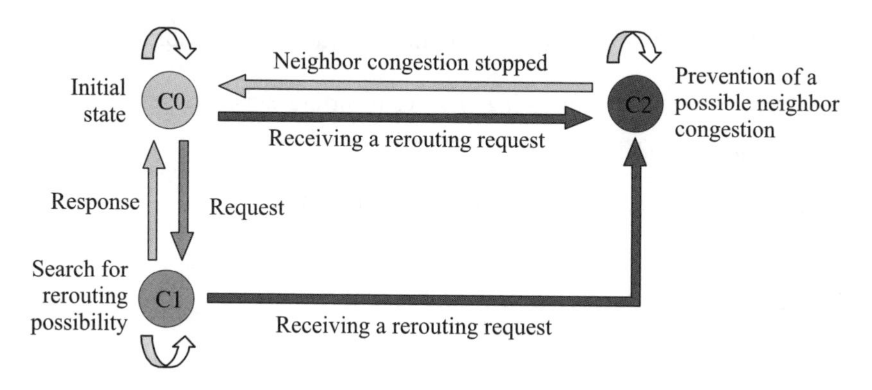

Figure 8.7. *Cooperative behavior of model 2*

"C0" represents the agent's state when no risk of congestion has been reported from a neighbor. "C1" describes the state of an agent receiving a request asking for its rerouting capacity. If the agent finds a rerouting possibility, it notifies the initiating agent by specifying the types of traffic that it can reroute. "C2" is reached following the reception of an urgent traffic redirection call which launches rerouting if possible. "C2" can be reached from "C0" or "C1" depending on whether the congested agent reaches its "L2" alert level before of after information retrieval. Cooperative behavior is explained with the following five rules.

COOP1: IF (C0) AND (receiving a request for information) THEN:

– transition to (C1);

– processing request for verification of rerouting possibilities.

COOP2: IF (C0) AND (receiving a rerouting request message) THEN:

– transition to (C2);

– modification of routing algorithm to transmit traffic to available connections.

COOP3: IF (C1) AND (end of request processing) THEN:

– transition to (C0);

– send response.

COOP4: IF (C1) AND (receiving a rerouting request message) THEN:

– transition to (C2);

– modification of routing algorithm to transmit traffic to available connections.

COOP5: IF (C2) AND (receiving message reporting end of congestion) THEN:

– transition to (C0);

– initialization of routing table to default connections.

8.5. Simulation – platform, topology and results

The validation of our proposal integrating agents in a telecommunications network requires the use of a platform capable of simulating networks and agents. Our problem lies in the fact that this type of platform does not exist. Several platforms have been proposed in the field of telecommunications network simulations, such as NS-2 [NS2 04], J-Sim [JSI 03, JSI 04], ANTS [WET 98] and others. These simulators model purely network elements with, for some, an integration of the mobile agents, without anticipating the possibility of integrating functionalities which help the implementation of intelligent agents in the network nodes. On the other hand, in the artificial intelligence field, different platforms have been proposed such as, for example, DECAF [GRA 00], JAFMAS [JAF 04], DMAS [DMA 04] and JACK [JAC 02]. All these platforms have been designed for these purely agent simulations which simulate a simple agent network. To be able to simulate a telecommunications network with this type of platform we must model all network elements as agents [GAI 02, LEC 02].

Finally, it appears obvious that a new platform must be developed. We have chosen to implement an extension of the J-Sim network platform with agent functions – this choice is justified in the next section. Two other sections present the simulated topology and results respectively.

8.5.1. *Platform*

We have chosen to implement an extension of a network platform with agent functions instead of modeling the network elements as agents in an agent platform. This choice is based on the fact that the use of a network simulator offers us more realistic results than modeling the needed network elements in an agent platform. In addition, most network platforms implement standard techniques proposed in the field and are also open in order to integrate new mechanisms easily. This enables us to effortlessly test and compare our agent behavior with different types of networks without having to redefine all these techniques.

Among the network platforms, J-Sim was chosen as being the most open to platform extensions and which already implements a large number of standard mechanisms proposed in networks such as IntServ, DiffServ, MPLS [ROS 01] and active networks.

J-Sim is a 100% Java-based simulator based on the notion of autonomous components. J-Sim proposes INET, which is an abstract network model. This model defines a node as a group of components and implements the basic elements necessary for node specification and for its extension with new protocols and mechanisms.

We have integrated the notion of intelligence in J-Sim [MES 04] by implementing an agent component which has the characteristics and functions required to implement different types of agents. This agent may have different behavior types – deliberative or reactive – based on the goal of its integration within the network. It contains a database in which it stores its knowledge and the resources that it can offer and request from other agents in its domain. The agent is able to communicate with the other network agents throughout the network infrastructure defined in J-Sim. However, this communication is not limited to a simple message transfer but can go as far as the definition of communication protocols such as Contract-Net [SMI 80]. In order to cooperate, the agents need to understand each other; this has required the specification of a communication language based on the theory of *speech acts*.

8.5.2. *Topology and configuration*

In all the simulations presented in this chapter we have adopted the topology described in Figure 8.8. This topology is broken down into three sub-networks N11, N12 and N13 where N11 contains H1 and H3 sources, N12 is the transmitting DiffServ network and N13 contains destinations H6 and H7.

In a DiffServ network, N1 and N2 represent the nodes which control flows and mark packets with a DSCP corresponding to a specific PHB. N4 and N5 are the last DiffServ nodes receiving packets before they leave the domain, N8 and N9 process the packets within the DiffServ domain following the requested PHB. These different nodes are connected by links weighted by costs which represent the load added to routing table calculations. These costs are modified by the agent according to the network load to modify the choice of the output link during the routing decision.

The network is configured as follows:

– the nodes are interconnected by 6 Mbps bandwidth connections except for the connection between N8 and N4 which has only 3 Mbps designed to provoke congestion;

– costs of all the connections are initialized to 1 except for N0-N9 and N2-N5, which are 2 and 4 respectively in order to force the choice of N0-N8 and N2-N8 as default links to reach the destination network.

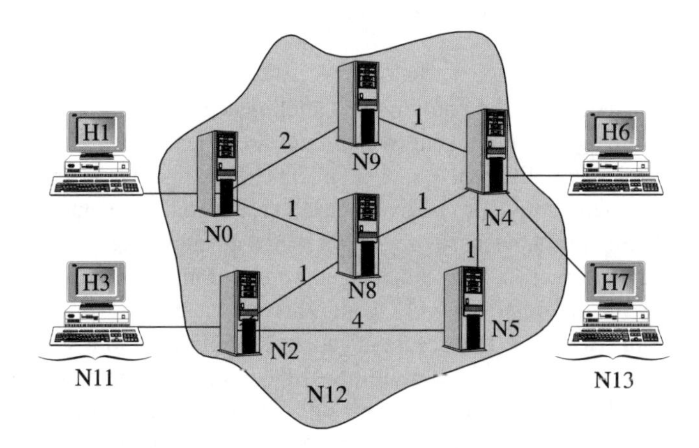

Figure 8.8. *Simulated topology*

The simulation parameters are:

– the simulation duration is 150 seconds;

– the delay of propagation between connections is configured at 0.5 seconds;

– the H1 and H3 sources respectively transmit EF and AF traffics 10 seconds after the start of the simulation;

– the EF packet size is 1,000 bytes;

– the interpacket interval of EF traffic is 0.013 seconds;

– the AF packet size is 1,000 bytes;

– the interpacket interval of AF traffic is 0.01 seconds.

8.5.3. *Simulation results*

The topology described in section 8.5.2 was simulated with different types of nodes. In the first simulation, the network is a *best effort* type network which does not integrate either DiffServ or agent. In the second simulation, the nodes are DiffServ nodes but do not integrate agents. In the third simulation, the agents are implemented in the nodes and adopt the first behavioral model described in section 8.4.3.1 and in the fourth simulation, the agents have the second behavior described in section 8.4.3.2. The results of these simulations are explained in the next sections and have the following interpretations:

– line 1 (dotted black) represents EF traffic sent by H1;

– line 2 (solid black) represents EF traffic received by H6;

– the difference between black lines represents EF traffic loss rate;

– line 3 (dotted gray) represents AF traffic sent by H2;

– line 4 (alternately dotted or solid gray) represents AF traffic received by H7;

– the difference between gray lines represents AF traffic loss rate.

8.5.3.1. *Simulation no. 1*

In the first simulation, we have chosen to show the behavior of a *best effort* network in order to compare it with the network behavior after DiffServ implementation and then integration of agents. This enables us to have an idea of the advantage of DiffServ networks with the assurance of the notion of QoS and of the impact of agent implementation in this type of network. The *best effort* network has no guaranteed QoS for the different types of traffic.

Figure 8.9 shows large losses in EF and AF traffics represented by the difference between 1/2 and 3/4 lines respectively.

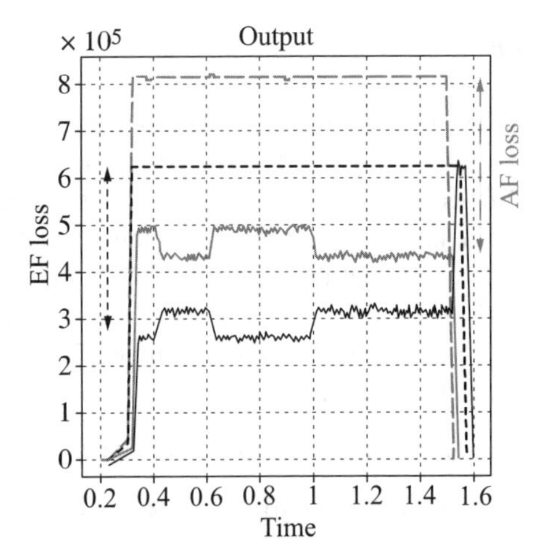

Figure 8.9. *Best effort network behavior*

8.5.3.2. *Simulation no. 2*

In the second simulation, we have chosen to show the behavior of a DiffServ network before the integration of agents in order to compare the network's global behavior with and without agents and show the improvement they introduced.

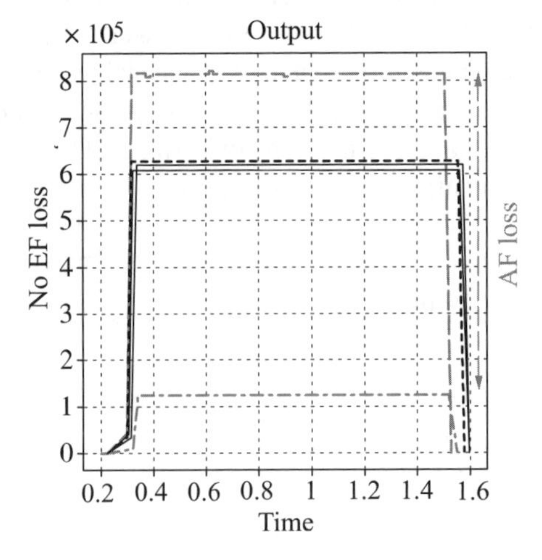

Figure 8.10. *Behavior of the DiffServ network before agent integration*

Figure 8.10 shows that after the implementation of the DiffServ technique, EF traffic requiring a large QoS was protected from losses. On the other hand, AF traffic, which has less priority than EF, experienced a very high loss rate. AF packets have been lost because of lack of resources in the N8-N4 connection since the route used by default by both types of traffic goes through N8 (cheapest route) whereas other routes, which are more expensive but with less load, are available.

8.5.3.3. *Simulation no. 3*

In this simulation, an agent is integrated in each network node with the objective of preventing congestion and minimizing the loss rate of AF packets while transmitting specific packets to nodes which are less loaded than the node at risk of congestion. This agent implements the behavioral "model 1", which is described in section 8.4.3.1.

Figure 8.11 shows a significant improvement in the loss rate of AF packets. Almost all AF traffic reaches destination – but obviously with a higher cost than if the cheapest route was taken – except for some loss caused by the queues' filling period.

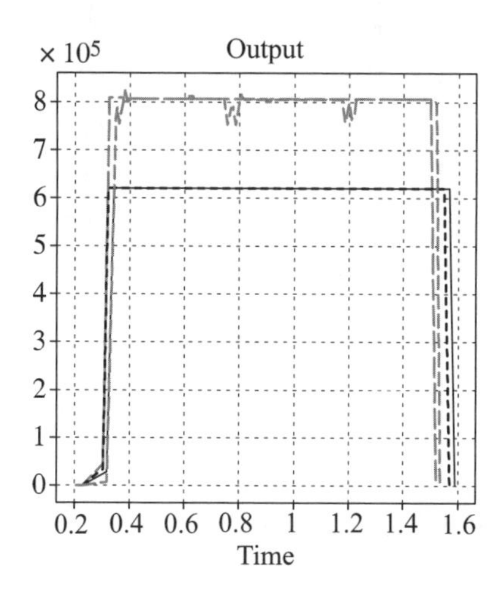

Figure 8.11. *Behavior of a DiffServ network integrating agents*

8.5.3.4. *Simulation no. 4*

In the fourth simulation, an agent is integrated in each network node with the same objective as the third simulation which consists of preventing congestion and minimizing the loss rate of AF packets while transmitting some traffic towards nodes which are less loaded than the node at risk of congestion, except that this agent implements the behavioral "model 2" described in section 8.4.3.2.

In the simulations presented in this section, the sources transmit after 40 seconds from the start of the simulation and not after 20 seconds as with the previous ones. This is because in this behavioral model, the agents need to initialize their resources before receiving traffic.

These simulations process the different following case studies corresponding to the agent's different possible transitions: (1) termination of information retrieval before going to state "P2" and (2) interruption of information retrieval and dispatch of rerouting request to all neighbors. These two case studies have the same network behavior seen by the destinations but differ in the distribution of network load between the different nodes.

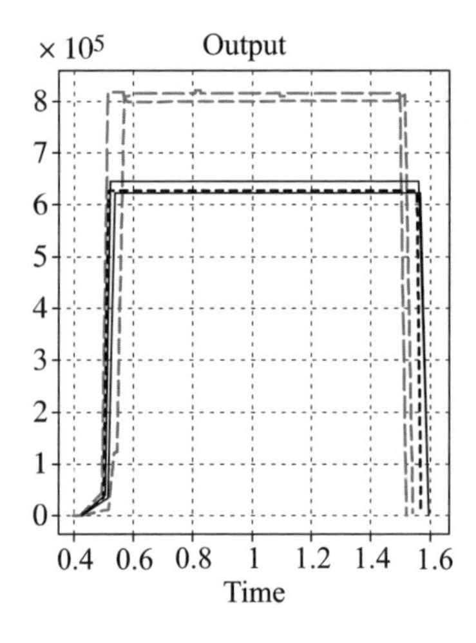

Figure 8.12. *Behavior of DiffServ network integrating agents with model 2 behavior*

In the simulation corresponding to Figure 8.12, AF packet loss rate seen by destinations is brought down to zero, which represents a significant improvement compared to previous simulations. In this simulation, both sources transmit simultaneously, which involves an information retrieval interruption and a dispatch of a rerouting request to all neighbors. This results in a reaction from all surrounding nodes with rerouting capacity, which modifies the routing of most of the traffic and makes this route the cheapest one going through the practically empty N8 after the detection of a congestion risk (Figure 8.13). Figure 8.13 shows EF (gray dotted line) and AF (black dotted line) traffic rate going through the N8-N4 connection and the total load of this connection (black solid line). Both types of traffic do not go through N8-N4 after congestion has been detected.

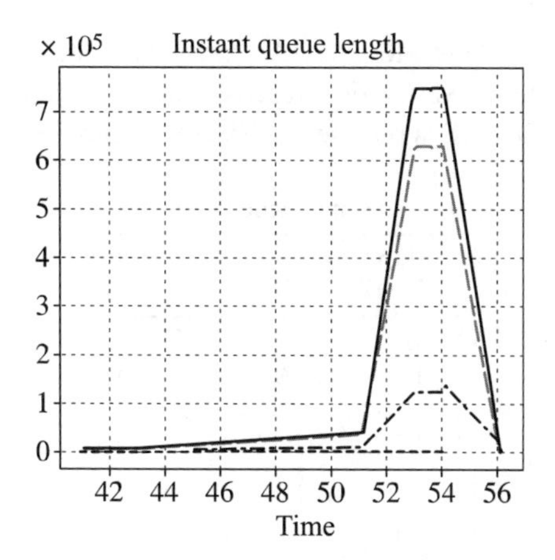

Figure 8.13. *Load rate of N8-N4 connection*

If the sources increasingly saturate the N8-N4 bandwidth by transmitting, for example, with a difference of at least 6 seconds (termination duration of information retrieval), the distribution of the traffic load in the network will be smarter by rerouting certain traffic (Figure 8.15). In this case, the loss rate seen by destinations remains zero as shown in Figure 8.14.

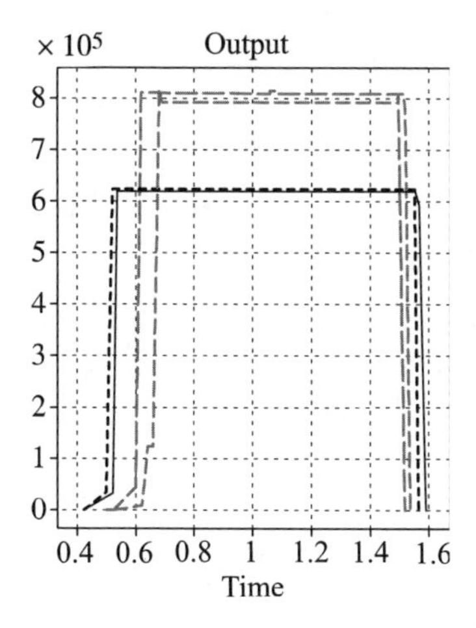

Figure 8.14. *Model 2 agent behavior without interruption of state P2*

The difference with the previous simulation is that EF traffic continues to use the default route (black dotted line), whereas AF is rerouted (end of gray line). In this simulation the agent has chosen to move AF traffic and not EF. This choice is based on the fact that the rate of AF traffic exceeds bandwidth capacities which means a loss of AF packets in case EF is rerouted to N9 (see Figures 8.16 and 8.17).

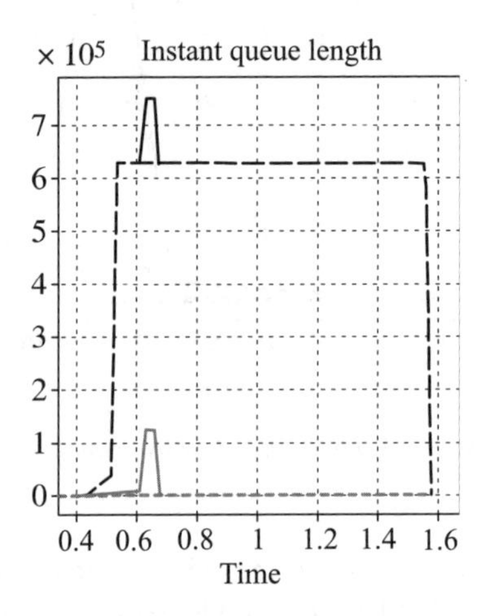

Figure 8.15. *Load rate of N8-N4 connection without interruption of P2*

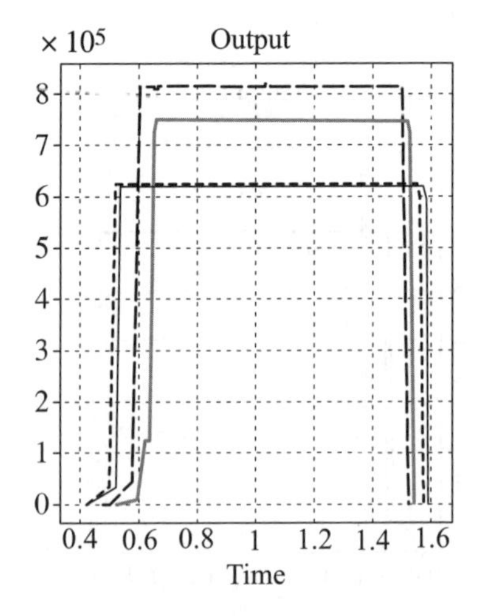

Figure 8.16. *Loss of AF packets if the routing decision was for EF*

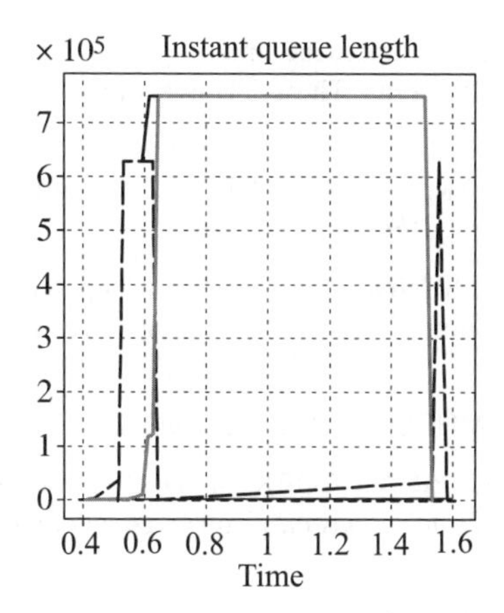

Figure 8.17. *Load rate of N8-N4 connection with routing decision for EF*

8.6. Conclusion

This chapter presented an MAS implementation in a DiffServ network with the aim of improving the performance of this type of network and to provide users with the best QoS. The effectiveness of this proposal had to be proven by simulations, which required the use of a platform able to simulate MAS and network elements simultaneously. Since this type of network does not exist, this chapter has proposed an extension of the Javasim network platform to integrate the notion of intelligence.

Two agent behavioral models have been implemented in the new Javasim extension. The results show a noticeable improvement in the global behavior of the DiffServ network, especially in the prevention of congestion which in turn implies the prevention of loss of packets requiring a QoS.

8.7. Bibliography

[APP 94] APPLEBY S., STEWARD S., "Mobile Software Agents for Control Distributed Systems based on Principles of Social Insect Behavior", *Proceedings from ICCS'94*, Singapore, 1994.

[BAZ 99] BAZZAN A., WAHLE J., KLUGL F., "Agents in Traffic Modeling – From Reactive to Social Behaviour", *Proceedings from KI'99*, Bonn, p. 303-307, September 1999.

[BLA 98] BLAKE S., BLACK D., CARLSON M., DAVIES E., WANG Z., WEISS W., "Architecture for Differentiated Services", *RFC2475*, December 1998.

[BOU 98] BOUKHATEM N., L'approche multi-agent pour un contrôle de congestion adaptatif de réseaux ATM, PhD Thesis, University of Versailles Saint Quentin-en-Yvelines, 1998.

[BRA 94] BRADEN R., CLARK D., SHENKER S., "Integrated Services in the Internet Architecture: An Overview", *RFC1633*, June 1994.

[BUR 92] BURMEISTER B., SUNDERMEYER K., "Cooperative Problem Solving Guided by Intentions and Perception", Werner, Demazeau (ed.), *Decentralized AI, 3*, Elsevier Science, 1992.

[DEM 91] DEMAZEAU Y., MÜLLER J.P., *Decentralized AI, 2*, Elsevier Science, 1991.

[DUR 89] DURFEE E., LESSER V., CORKILL D., "Trends in Cooperative Distributed Problem Solving", *IEEE Transactions on Knowledge and Data Engineering*, vol. KDE-1, p. 63-83, March 1989.

[ETZ 94] ETZIONI O., WELD D., *A Softbot-Based Interface to the Internet*, 1994.

[FER 97] FERBER J., *Les systèmes multi-agents: vers une intelligence collective*, InterEditions, Paris, 1997.

[GAI 02] GAÏTI D., MERGHEM L., "Active Network Modelling and Simulation: A Behavioral Approach", *Proceedings from SmartNet'2002,* Finland, p. 19-36, April 2002.

[GAL 88] GALLIER J.R., A theorical framework for computer models of cooperative dialogue, acknowledging multi-agent conflict, PhD Thesis, Open University, UK, 1988.

[GAR 03] GARNEAU T., DELISLE S., "A New General, Flexible and Java-based Software Development Tool for Multiagent Systems", *Proceedings from ISE2003 International Symposium on Information Systems and Engineering,* Montreal, Canada, July 20-25 2003, p. 22-29

[GRA 00] GRAHAM J., DECKER K., MERSIC M., *DECAF – A Flexible Multi-Agent System Architecture*, Kluwer Academic Publishing, 2000.

[GUE 96] GUESSOUM Z., Un environnement opérationnel de conception et de réalisation de systèmes multi-agents, PhD Thesis, Paris 6 University, 1996.

[HEI 99] HEINANEN J., BAKER F., WEISS W., WROCLAWSKI J., "Assured Forwarding PHB Group", *RFC2597*, June 1999.

[HUS 00] HUSTON G., "Next Steps for the IP QoS Architecture", *RFC2990*, November 2000.

[JAC 02] JACK TEAM, "JACK Intelligent Agent user guide", Release 3.5, available at http://www.jacksoftware.com/shared/demosNdocs/JACK_Teams_Manual.pdf, March 5 2002.

[JAC 99] JACOBSON V., NICHOLS K., PODURI K., "An Expedited Forwarding PHB", *RFC 2598*, June 1999.

[JAF 04] JAFMAS GROUP, available at http://www.ececs.uc.edu/~abaker/JAFMAS, 2004.

[JSI 03] J-SIM GROUP, *Creating Simulation Scenarios*, Ohio State University, October 2003.

[JSI 04] J-SIM GROUP, available at www.J-Sim.org, 2004.

[LEC 02] LECARPENTIER H., GAITI D., MERGHEM L., "Telecommunication Network Modeling Using an Organized Group of Agents", *Proceedings from ICTSM 10*, Monterey, USA, p. 783-793, October 2002.

[MER 02] MERGHEM L., LECARPENTIER H., GAITI D., "Behavioral Multi-agent Simulation of an Active Telecommunication Network", *Stairs'2002*, Lyon, p. 217-226, July 2002.

[MES 03a] MESKAOUI H., GAÏTI D., KABALAN K., "Implementation of a Multi-agent System Within a DiffServ Network to Improve its Performance", *Proceedings from ISE'2003*, Montreal, Canada, July 2003.

[MES 03b] MESKAOUI H., GAÏTI D., KABALAN K., "DiffServ Network Control using a Behavioral Multi-agent System", *Proceedings from Netcon'2003*, Muscat, Oman, October 2003.

[MES 04] MESKAOUI H., GAÏTI D., "Intelligent Features Within the J-Sim Simulation Environment", *Proceedings from Intellcom'2004*, Thailand, October 2004.

[MIN 94] MINSKY M., "A Conversation with Marvin Minsky", *Communication of the ACM*, 1994.

[MOU 98] MOUKAS A., CHANDRINOS K., MAES P., "Trafficopter: A Distributed Collection System for Traffic Information", *Proceedings from CIA'98*, Paris, p. 34-43, July 1998.

[NS2 04] NS-2 GROUP, available at www.isi.edu/nsnam/ns, 2004.

[ROS 01] ROSEN E., VISWANATHAN A., CALLON R., "Multiprotocol Label Switching Architecture", *RFC 3031*, January 2001.

[SHO 93] SHOHAM Y., *Agent Oriented Programming, Artificial Intelligence*, 1993.

[SIG 02] SIGEL E., DENBY B., HEGARAT-MASCALE S.L., "Application of ant colony optimization to adaptive routing in LEO telecommunication satellite network", *Annals of Telecommunications*, vol. 57, p. 520-539, May-June 2002.

[SMI 80] SMITH R.G., "The Contract-Net Protocol: High-Level Communication and Control in a Distributed Problem Solver", *IEEE Transaction on Computers*, vol. 29, no. 12, p. 1104-1113, 1980.

[THA 99] THANH D.V., STEENSEN S., AUDESTAD J., "Mobile Agents for Mobile Users", *Proceedings from Smartnet'99*, Thailand, November 1999.

[WET 98] WETHERALL D., GUTTAG J., TENNENHOUSE D., "Ants: A Toolkit for Building and Dynamically Deploying Network Protocols", *IEEE OPENARCH'98*, San Francisco, April 1998.

[WRO 97] WROCLAWSKI J., "The Use of RSVP with IETF Integrated Services", RFC 2210, September 1997.

Chapter 9

Intelligent Agent Control Simulation in a Telecommunications Network

9.1. Introduction

Telecommunications networks are in a transitional phase where commercial telecommunications or computer manufacturers not only merge their basic concepts, but are also working towards a common goal: completely cooperative communication systems for a dynamic context adaptation and task automation.

Potential application users requiring a specific quality of service are increasingly demanding and numerous. In addition, new types of networks are networks of networks (wired, fiber optic, radio cell, satellite, Wi-Fi, etc.) which are managed by different service providers and therefore managing these networks is a problem (invoicing, guarantee of quality of service, negotiation between operators and service providers).

In order to respond to new user expectations (quality of service, high throughput, mobility, reliability, etc.), operators and service providers oversize. Nevertheless, it is a medium-term and costly solution which does not consider end-to-end quality of service problems.

Several research studies are focused on the use of intelligent software agents for management and network control as well as for negotiation between operators, maintenance and MMIs (*man-machine interfaces*). These studies use different types

---

Chapter written by Hugues LECARPENTIER.

of agent architectures, which are difficult to model with traditional network simulators.

We propose integrating a generic agent model in a telecommunications network simulator and using it as a basis for all agents to be simulated, regardless of their task. In this way, we can simulate several types of agents simultaneously and enable them to communicate in a simple manner. This tool will eventually enable us to simulate network behavior by using software agents in numerous contexts (control, MMI, negotiation, etc.) and networks which group several operators using different types of agents.

The multi-agent technology is a recent innovation which comes from several research fields: symbolic artificial intelligence, control theory and distributed artificial intelligence. The rapid development of this technology is explained by the demand for new solutions and new mechanisms for the modeling and resolution of complex problems.

This technology is increasingly turning out to be a key technology as computer systems are becoming even more distributed, interconnected and open. In these types of environments, the capacity of software agents to plan and perceive their actions and goals, to cooperate and negotiate autonomously with others, and to respond in a flexible and intelligent manner to dynamic and unpredictable situations, significantly improves the quality and performance of these software systems.

Telecommunications infrastructures seem to be a natural application for software agents [HAY 99]. A particularly important aspect in the relation between telecommunications networks and multi-agent systems is that agents communicate through a standard and universal language, whereas telecommunications sub-systems are highly heterogenous and complex. This multi-agent system property makes it possible for us to consider end-to-end quality management, regardless of the networks crossed.

First, we will look at studies on network control and management by intelligent software agents. Two intelligent simulation examples are then explained, followed by a detailed discussion on the simulation tool used. Section 9.5 is dedicated to the chosen intelligent agent architecture and in section 9.6 we present an illustration of the capacities achieved with the simulator.

9.2. Network management and control by intelligent software agents

There is a large number of network management and control projects using intelligent software agents. Most of the studies focus on ATM networks and mobile agents. For a more detailed review see [HAY 99].

9.2.1. *Agent-based admission control*

Gibney and Jennings [GIB 97] propose the use of an approach based on market mechanisms (*market-based*) for admission control in ATM networks. Intelligent agents are used for resource allocation in a broadband environment. These studies emphasize efficiency and reliability of multi-agent systems in distributed and dynamic environments. They start with the concept that ATM networks are very high throughput networks and therefore data relative to network management quickly becomes obsolete. They propose to use economy theories to proactively pair the network's dynamic nature with local control activity. They demonstrate how market-based control provides better performances than traditional methods.

9.2.2. *Project Tele-MACS*

Project Tele-MACS [HAY 98] uses a multi-agent control approach for ATM network management. A multilayer agent architecture was developed. Reactive control agents and proactive planning agents are used to avoid congestion and maintain network infrastructures. Admission control is also addressed. Project Tele-MACS is a group of multi-agent systems organized in layers where each layer is responsible for controlling network infrastructures with an inherent skill level. The use of a multi-agent control system is totally relevant to broadband network constraints, i.e. the need for intelligent control and intelligence movement from switches to edges.

9.2.3. *Project Hybrid*

Project Hybrid [SOM 96] is based on a three-layer multi-agent architecture for ATM network management. The architecture is based on an autonomous agent hierarchy for network control based on local decision making capabilities. This architecture has control distribution, hierarchy and autonomous control properties.

The hierarchy between the three layers is distributed according to geographical zones: local, regional and national. Each layer is responsible for a specific geographical zone. Hybrid uses the FCAPS (*fault*, *configuration*, *accounting*, *performance and security functional areas*) model [URL 1] and negotiation protocols to resolve conflicts in the system.

9.2.4. *Route selection by mobile agents*

A mobile agent application for congestion control in switched circuits is presented in [APP 94]. These studies demonstrate the potential of software agents for autonomous network management. In fact, distributed network control requires the use of complex coordination strategies. The authors use a large number of agents for controlling routing in a 30 node SDH (*synchronous digital hierarchy*) network. This approach provides a high control strength.

9.2.5. *Cooperative mobile agents for network mapping*

The use of mobile agents for the discovery of information on a network's connectivity is presented in [MIN 98]. The agents build a map of the network containing configurations for all the nodes and their links to facilitate dynamic routing decisions. This approach is particularly efficient when centralized solutions are difficult to implement because of the dynamic character of the network.

Each agent uses a three step control cycle. These three steps are: discovering the node's links where the agent is located, collecting information from other agents present on the node and choosing the next node to reallocate. Several movement strategies have been used during simulations on a 250 node network.

Results show that the best results are obtained with the use of highly cooperative agents. It is important to note that these results also show the existence of an optimum number of agents to improve network mapping. If this number is exceeded, performance drops. An online demonstration is available on the Internet [URL 2].

9.2.6. *Project MAGNA*

This project proposes an approach based on mobile agents to improve and extend TINA-C capacities [URL 3]. The objective of this project is the development of a generic architecture responding to future communication network requirements [MAG 96]. MAGNA has emphasized the efficiency of mobile agents to solve scaling problems with services in intelligent networks.

9.3. Simulating the behavior of intelligent agents in a communication network

Current network simulators are not able to model and simulate intelligent software agent behaviors. Relevant simulation experiments are rare and focused on a specific problem. Two solutions are used:

– development of a network/multi-agent simulator based on a multi-agent simulation platform;

– development of a network/multi-agent simulator based on a network simulation tool.

9.3.1. *Simulation of behavioral quality of service network control*

Works presented in Chapters 6 and 7 and in [MER 03] use the oRis multi-agent platform [URL 4] to model and simulate a network with quality of service. A behavioral multi-agent system controls each network router. Agent/router pairs adapt their behavior according to the flow to process, the filling of their queues and the behaviors of their neighbors. Several types of behaviors have been defined according to whether they act on scheduling, queue management, routing or message forwarding to other nodes.

The use of a multi-agent platform has required the implementation of all network elements and traffic generation methods. On the other hand, all the properties of multi-agent systems such as message management were already in place. This tool is adapted to types of simulations formed and proposes new graphical visualization components.

The network model developed is very focused and changes needed to make it usable for new types of networks are significant. This solution is significant if the type of network simulated is always the same. It makes it possible to simulate the control of different types of agents by simplifying their implementation.

9.3.2. *Intelligent control simulation of a DiffServ network*

Meskaoui proposes in Chapter 8 and in [MES 03] to use the simulation tool from J-Sim networks [URL 5] to simulate the control of a multi-agent system in a DiffServ network with quality of service [URL 6]. The J-Sim simulator is extended by the addition of software agents for controlling the router and negotiation protocols. These extensions are easily achieved because of the autonomous component architecture used by J-Sim (see section 9.4).

The simulation of different types of networks is very simple because the simulator contains complete and easily adaptable libraries. On the other hand, the agents used are dedicated to DiffServ router control and we may have to implement entirely new architectures.

9.3.3. *Comparison and choice of a platform*

Both of these studies offer advantages and drawbacks (see Table 9.1). However, the complexity and diversity of communication networks clearly favor J-Sim. In addition, its component-based approach makes it possible to simply integrate agent properties and cooperation, interaction and communication protocols. J-Sim seems therefore to be an excellent candidate for the design of a tool for control simulation by intelligent agents. J-Sim is discussed in more detail in section 9.4.

	Addition of multi-agent properties	**Addition of network properties**	**Multi-agent specialist use**	**Network specialist use**
oRis	Very simple	Complex	Simple	Complex
J-Sim	Simple	Very simple	Relatively simple	Very simple

Table 9.1. *Comparison between oRis and J-Sim*

9.4. Detailed simulator presentation

J-Sim is a simulation kernel designed with autonomous component architecture and entirely developed in Java. J-Sim component behaviors are defined in terms of agreements and can be individually designed, tested and deployed in a software system. The system to simulate may be made up of components in the same way that we put together chips to make a circuit. In addition, we can add new components during execution and a graphical interface is available (see Figure 9.1). These characteristics make J-Sim a portable, progressive and reusable environment.

With this architecture we can compose different simulation scenarios based on the type of network we wish to simulate, from a set of basic components and components developed by the user.

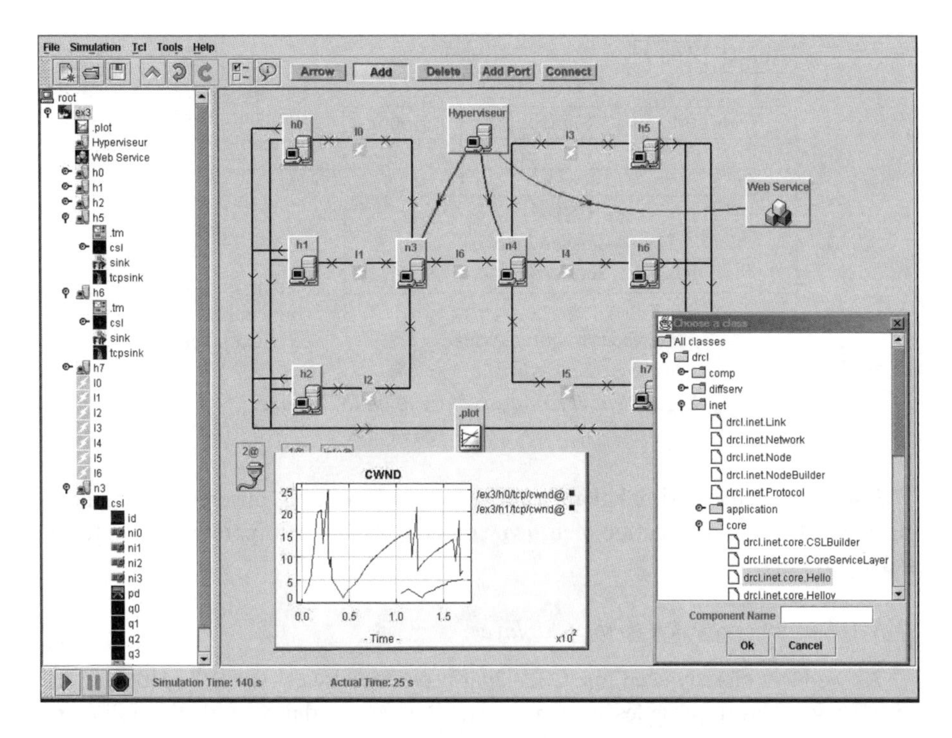

Figure 9.1. *J-Sim graphical interface*

9.4.1. *Structure of an INET node*

J-Sim proposes a network abstraction model over the simulation kernel: INET.

The internal node structure (router or computer) is made up of two layers: CSL (*core service layer*) and ULP (*upper layer protocol*) that we can see in Figure 9.2. The CSL layer contains network, connection and physical layers, and provides a set of predefined services for ULP layer modules. The ULP layer contains transport, signaling and application protocols.

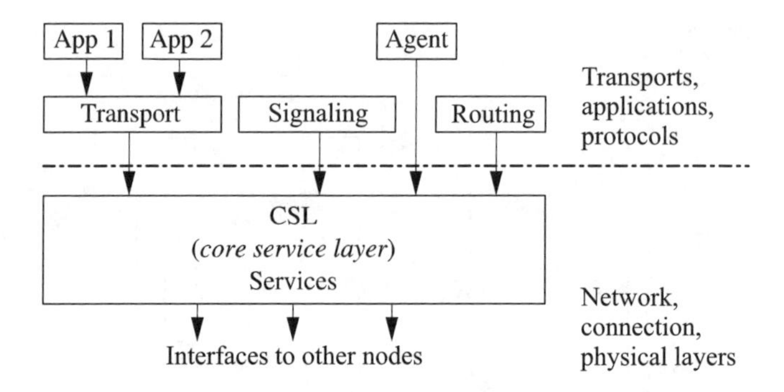

Figure 9.2. *Internal structure of an INET node*

Users can create simulation scenarios by connecting different components together. They can also define their own components and incorporate new protocols and algorithms.

9.4.1.1. *Services provided by the CSL layer*

As we have already seen, the CSL layer provides services to ULP modules. The basic service layer provides customers (ULP layer modules) with the following services:

– data transmission;

– node identity management;

– routing table management;

– neighborhood management;

– packet filtering.

These services are defined in terms of agreements. A protocol module using basic services must have a port connected to a specific service contract or agreement (see Figure 9.3). The protocol module is connected to the CSL layer by linking one of its ports to the corresponding CSL layer port. These operations are executed during simulation tool design.

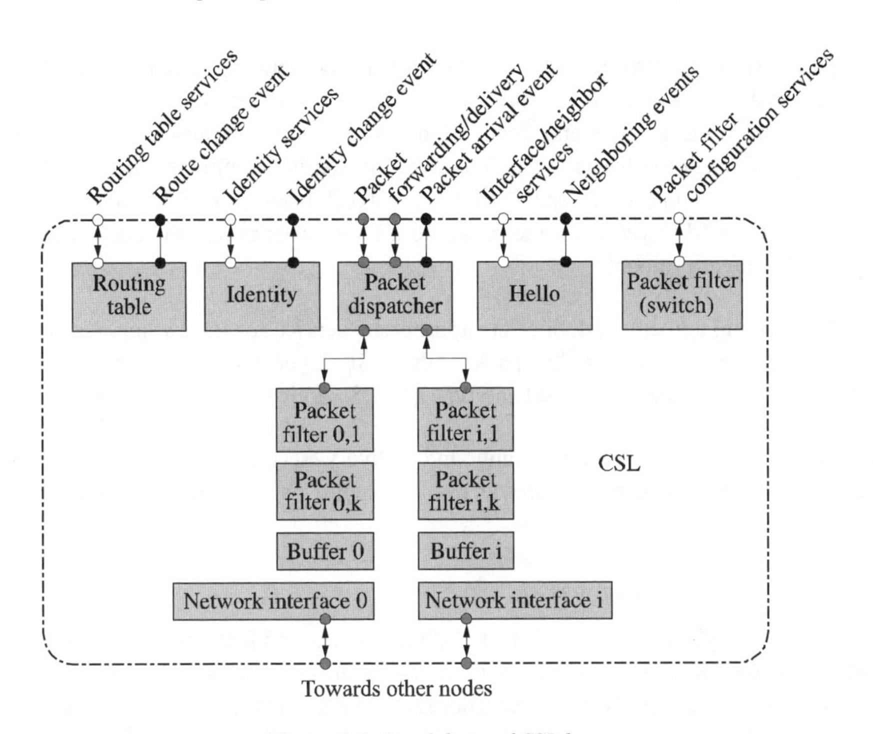

Figure 9.3. *Breakdown of CSL layer*

As described in Figure 9.2, a basic INET node is made up of the CSL layer and one or more transport and application protocol layers. We should mention that these transport and application protocols are located on the host nodes and not on network routers.

9.4.1.2. *Data transmission service*

The data transmission service consists of two agreements, one for each data flow direction (in and out).

The forwarding agreement defines the packet forwarding service provided by the CSL layer to ULP layer protocols. To send a packet, a ULP layer protocol must provide a destination address, TTL (time to live, i.e. number of hops before the packet gets destroyed), router alert flag (indicating if the packet must be intercepted by intermediate routers or not) and TOS (type of service) that the packet must receive. In the current implementation, INET only uses the first bit of TOS fields to separate data packets from control packets. This process can be refined to provide more options to ULP layer protocols.

The CSL layer provides three data forwarding options: default dispatching, explicit multicast and exclusive broadcast. For default dispatching, the CSL layer looks at the routing table and destination address and determines which output interface the packet will use to be delivered. For explicit multicast, the CSL layer sends packets to interfaces specified by the ULP layer protocol. For exclusive broadcast, the CSL layer sends the packet to all interfaces except the ones specified by the ULP layer protocol.

The arriving agreement defines the appropriate service for forwarding a packet to a ULP layer protocol. For this packet, the CSL layer provides sender address, interface where the packet arrived and type of TOS service.

In addition to packet forwarding and delivery services, the CSL layer can perceive and distribute packet arrival events when a packet arrives on a lower or higher port.

9.4.1.3. *Node identification service*

The CSL layer maintains a list of node addresses or identities known by other network nodes. In INET, a monopoint address is an identity owned by at least one network node, whereas a multicast address is an identity by which at least one network node can be identified. A ULP layer protocol can look at and/or configure node identities through both service contracts detailed below.

The identity consultation agreement defines the service that a ULP layer protocol uses to research node identities. Three types of consultation services are available: default identity consultation, identity search and all identities search. An identity is set as default identity for each node. In the default identity consultation service, the CSL layer sends the node's default identity. In identity search service, the requester specifies the identity researched and the CSL layer sends a true Boolean value if this identity is one of its own (false if not). In the search service of all identities service, the CSL layer sends the list of all node identities.

The identity configuration agreement defines the node's identities configuration services, i.e. add/delete an identity. In addition, each identity can be associated with a countdown. When it expires, the associated identity disappears. This service also makes it possible to adjust or consult the timer.

An identity change event (all modifications are reported) can also be reported by the CSL layer.

9.4.1.4. *Routing service*

The CSL layer of each node updates a routing table. For each packet to be processed, the appropriate interface is researched according to a source address, destination address and interface on which the packet arrived. A ULP layer protocol can review and/or configure the routing table through the following service agreements:

– *route search:* in this agreement the requester specifies the source and destination addresses and the arrival interface. The CSL layer then sends the forwarding interface where the packet must be sent;

– *route configuration:* this agreement defines the configuration services of the routing table. In addition to being able to add or delete table entries, we can also retrieve or modify an entry. Furthermore, each entry is associated with a timer and the entry is deleted when it ends. We can also adjust or review each timer;

– *route request:* the CSL layer defines a route request agreement when the route does not exist for a given packet. With this service, the node can add new entries to the routing table or use temporary entries.

The CSL layer sends an event when the routing table is modified.

9.4.1.5. *Neighborhood management service*

The CSL layer has a database holding information from all the node's interfaces. The information of an interface contains a local address, address of neighbors which can be joined by this interface, size of output buffer, bandwidth, etc. The CSL layer provides information consultation and neighborhood events notification services to ULP layer protocols (new neighbor discovered, existing neighbor disappeared).

9.4.1.6. *Packet filtering services*

Conceptually, packet filters are the CLS layer's extensible part. Each node interface must contain a bank of filters. When the CSL layer sends a packet over an output interface, the packet must go through a set of filters before being placed in the interface buffer. When a packet arrives over a node interface, it can also go through a group of filters. If different CSL layer implementations can have different filters, all implementations must provide this configuration service so that ULP layer modules can configure existing filters.

9.4.1.7. *CSL layer composition*

The CSL layer revolves around five main components providing the services described previously (see Figure 9.3). The description of each of these components provides a good understanding of actions that software agents can take.

The *identity* component manages the different node identities and provides identity management services for ULP layer protocols and for the other CSL layer components.

The *routing table* component manages routing table entries and provides routing services for ULP layer protocols and for other CSL layer components.

The *packet forwarding* component provides data transmission services to ULP layer protocols.

The *hello* component updates interface information and node neighbors. It also provides neighborhood management service for ULP layer protocols and other CSL layer components.

The *packet filter switcher* is the component with the most detail. It is possible to execute most of the operations to guarantee quality of service with this component: scheduling management, queue management, choice of algorithm. This component acts between a ULP layer protocol and the *packet filter* components contained in the group of nodes. It can select the configuration (active filters) to use for packet processing. Each node contains a bank of packet filters, i.e. a group of packet filters linked together constituting a node interface. Each filter can execute specific operations on packets or reject some packets according to specified parameters. The last filter from a bank of filters is the network interface. This interface represents the hardware peripheral introducing packets in the network. Packet filters can be configured by modules of the ULP layer (they are generally signaling protocols or intelligent agents) via a packet filtering service. Each filter is indexed by an identifier in the bank of filters. ULP layer modules send a request with the identifier of the filter to activate.

9.5. Software agent architecture

Several software agent architectures are described in other works [HAT 03, SUG 01, TRE 02]. Our objective is to propose an architecture dedicated to network control and management, but we also want this generic architecture to be usable in different case studies.

Software agent architectures are generally classified into four families:

– deliberative architectures: designed around an explicit world representation, they use symbolic reasoning techniques. They are also called cognitive architectures;

– reactive architectures: the simplest architectures, of type reflex, they are based on a perception/action model based on rules;

– hybrid architectures: combination of reactive and deliberative architectures, these are the most flexible architectures;

– layer architectures: architecture based on a hierarchical organization.

As with numerous domains relying on time, the most significant architectures are hybrid architectures. They enable the agent to react immediately in situations requiring quick decisions. On the other hand, the agent can act in a reasoned way if the situation does not need a quick response but instead a more efficient one.

Our architecture is based on that proposed in [TSA 00] offering a basic architecture for software agents in telecommunications networks. This architecture adapts to all types of agents (mobile, fixed, interface, etc.) and it is also available for providing flexibility for software agents located over infrastructures controlling and managing the network.

The architecture chosen is presented in Figure 9.4 and the five main components are explained below. In cases where the agent has a simple behavior, advanced functions are deactivated. On the other hand, the architecture can be enhanced with a learning module if needed.

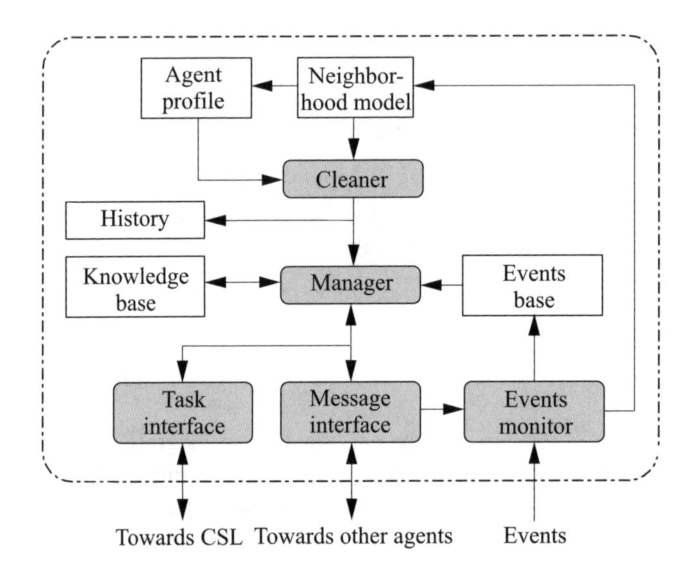

Figure 9.4. *Structure of a software agent*

9.5.1. *Events monitor*

This component is responsible for the detection and management of events internal and external to the node. The monitor receives all events distributed by the CSL: packet arrival, packet rejection, modification of routing table, modification of identities, modification of packet filters, threshold overflow in buffers, interface failure, etc. The monitor also perceives external events such as alert messages from the operator and certain messages from other network agents through the message interface.

The monitor classifies events in a database that it keeps updated and which is then interrogated by the agent manager. It also updates the neighborhood model and agent profile. The agent profile is a set of attributes such as agent name, type, role, capabilities and goals. These attributes are initialized during agent development and are then updated by the events monitor. They help to guide agent behavior during task planning, decision-making or even during interactions with surrounding agents.

The neighborhood model groups profiles of surrounding agents, i.e. those directly available through one of the network interfaces.

9.5.2. *Cleaner*

The cleaner module assembles events, tasks, agent profile and neighborhood model to provide the operator with consistent information. It cleans by deleting redundant or useless information. The operator can then always have relevant information on the node state from a temporal and operational standpoint.

This module is vital for optimum software agent operation. The work done by the cleaner is important. In fact, too much information will impede decision-making, but too little information will generate less efficient decisions. The cleaner is responsible for the information and its relevance.

9.5.3. *Message interface*

The message interface enables forwarding and arrival of messages. These messages are exchanged between software agents. The agents can use messages from their peers to update their global knowledge (network and other agents), but they can also inform, alert or interrogate other agents.

The messages exchanged use an agent communication language (ACL) proposed by FIPA (Foundation for Physical Intelligent Agents) [URL 7, URL 8]. FIPA ACL is the standard language for messages and requires coding, semantics and pragmatics of messages. The standard FIPA recommends common forms for conversations between agents by specifying interaction protocols, including simple query/response protocols, but also specific agent protocols such as contractual network and English and Dutch biddings.

FIPA ACL has 21 communication proceedings to express different types of messages. These communication proceedings can be classified into five groups: information transmission, information request, negotiation, action execution and error handling.

Each FIPA ACL message contains a group of elements presented in Table 9.2.

Element	Significance
Performative	Type of communication procedure
Sender	Message transmitter
Receiver	Message recipient
Reply-with	Participant in communication proceeding
Content	Content of message
Language	Content language
Encoding	Content encoding mode
Ontology	Ontology used to give content meaning
Protocol	Protocol controlling conversation
Conversation-id	Conversation identifier
In-Reply-to	Reference to message to which the agent responds
Reply-by	To force a delay for the response

Table 9.2. *Elements of a FIPA ACL message*

In our context an example of a message would resemble this:

```
(inform
    :sender Node A
    :receiver Node B
    :reply-with hypervisor
    :language KIF
    :ontology IP networks
    :content ( = (buffer1 threshold) (string exceeded))
    :conversation-id conversation01
)
```

9.5.4. *Task interface*

This module accomplishes control operations decided by the agent manager on the CSL layer, i.e. it acts directly on CSL layer components and can therefore control the routing table, filters and identifiers. This control is done through services provided by the CSL layer.

9.5.5. *Manager*

The management module makes the decisions and plans software agent actions. In order to do this, it uses a knowledge base. This knowledge base is made up of rules, properties, capabilities and any useful information for decision-making and plan production. The manager can then decide how and when to execute a given task.

It also manages the history in which it stores activated behaviors and observations. It also uses an events base to verify the efficiency of its actions and to understand future node reactions.

9.6. Illustration

In this section, we present an illustration emphasizing simulator possibilities. This example shows the simulator's capacity to simulate intelligent agents responsible for the control of quality of service for a given application. The chosen application is voice over IP, which is restrictive and simple to understand.

9.6.1. *Quality of service control for voice over IP*

In order to use voice over IP in a traditional IP network, we propose to combine each router with a software agent that will control data flow. The problem is how to pass voice flows through the network knowing that there already is substantial web traffic transmitting in this network. To guarantee sufficient voice quality, these flows are restrained: 100 ms delay, no packet loss and a jitter (gap between packets) of less than 20 ms. We presume that voice packets are marked. The agents will then choose the best scheduling and queue management algorithms, so that we do not lose these packets and guarantee the constraints.

The intelligent agent will obviously choose the algorithms which will put voice traffic in front, but also make comparisons between TTLs of several voice packets to serve the latest packets in priority.

The agent chooses algorithms enabling voice traffic to always be first in line and can then reject or delay web traffic. It can revert back to more traditional algorithms when voice traffic is no longer present or only slightly present in the queue. It follows routing and packet arrival events and acts on packet filter configuration according to queue filling.

If the agent predicts voice traffic congestion, it will inform the other agents so that they only send the traffic that must necessarily go through it. The agents receiving these messages modify their routing table (for a determined time) unless this is risky for their own state.

9.6.2. *Presentation of agents and routers used*

Figure 9.5 presents the agent and CSL layer of routers used. The agent listens to their routing, packet arrival and memory buffer threshold events. It can act on filter configuration and routing (choice of scheduling and queue management algorithms).

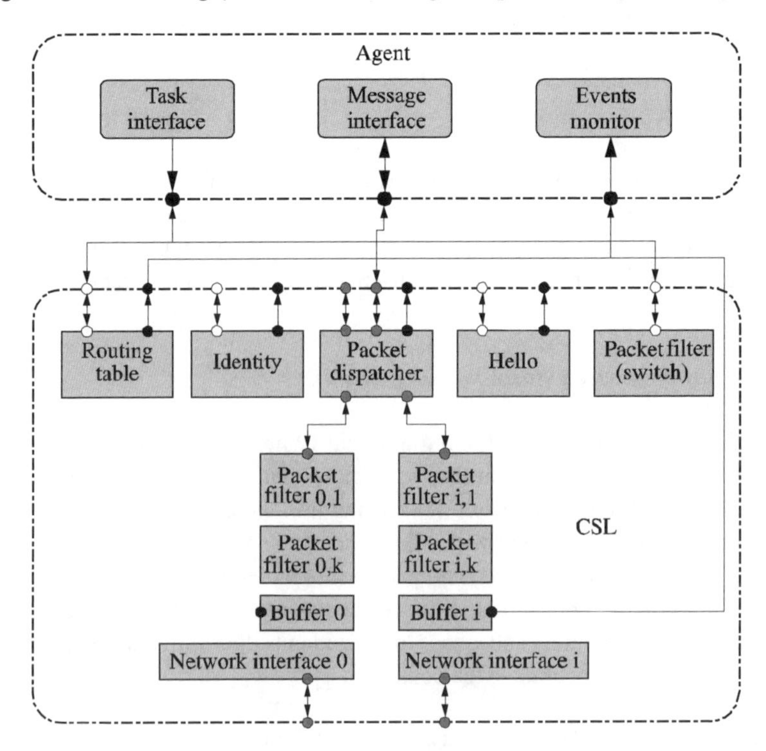

Figure 9.5. *Control of agent over router*

9.7. Conclusion

In this chapter, we have described the design of a tool capable of simulating control of intelligent agents in a network. The basis for this tool is a network simulator based on very simple components to simulate their progress. A dedicated agent architecture is proposed. Each of these agents can control network nodes via a set of services and event observers. The generic aspect of agent architecture gives this tool tremendous potential. It will indeed enable the simulation of different types of intelligent controls in different types of networks. A brief illustration was presented.

The growing number of works on control and network management by intelligent agents clearly demonstrates the necessity for an efficient simulation tool. Because of the cautiousness of telecommunications network players, simulation appears vital but traditional simulators are quickly found to be limited. Our simulator responds to a large need and enables us to face the huge infatuation for agent-based network technologies.

Our next objective is to simplify modeling and simulation processes in the intelligent agent/communications network context to enable users to concentrate on reasoning, interaction and learning aspects. Another objective is the definition of usable agent libraries from the generic architecture. These agents will be dedicated to types of networks or to specific problems.

9.8. Bibliography

[APP 94] APPLEBY S., STEWARD S., "Mobile software agents for control in telecommunications networks", *BT Technology Journal*, vol. 12, no. 2, p. 104-113, April 1994.

[GIB 97] GIBNEY M.A., JENNINGS N.R., "Market-Based Multi-Agent Systems for ATM Network Management", *Proceedings of 4th Communications Networks Symposium*, p. 62-65, Manchester, 1997.

[HAT 03] HATTORI M., OHSUGA A., HONIDEN S., "Planning mobile agent architecture for open network environment", *Systems and Computers in Japan*, vol. 33, no. 1, p. 58-66, January 2002.

[HAY 98] HAYZELDEN A., "Telecommunications Multi-Agent Control System (Tele-MACS)", *Proceedings of 13th European Conference on Artificial Intelligence*, p. 319-320, Brighton, UK, August 1998.

[HAY 99] HAYZELDEN A., BIGHAM J., *Software Agents for Future Communication*, Springer-Verlag, Berlin, May 1999.

[MAG 96] MAGEDANZ T., ROTHERMEL K., KRAUSE S., "Intelligent Agents: An Emerging Technology for Next Generation Telecommunications", *Proceedings of IEEE INFOCOM'96*, p. 464-472, San Francisco, March 1996.

[MER 03] MERGHEM L., GAÏTI D., PUJOLLE G., "On Using Multi-agent Systems in End to End Adaptive Monitoring", *Proceedings of 6th International Conference on Management of Multimedia Networks and Services*, p. 422-435, Belfast, September 2003.

[MES 03] MESKAOUI N., MERGHEM L., GAÏTI D., KABALAN K., "Diffserv Network Control Using a Behavioral Multi-Agent System", *Proceedings of Conference on Network Control and Engineering for QoS*, p. 51-63, Muscat, Oman, October 2003.

[MIN 98] MINAR N., KRAMER K.H., MAES P., "Cooperating Mobile Agents for Mapping Networks", *Proceedings of the First Hungarian National Conference on Agent Based Computing*, 1998.

[SOM 96] SOMERS F., "HYBRID: Intelligent Agents for Distributed ATM Network Management", *Proceedings of 12th European Conference on Artificial Intelligence, IATA Workshop*, Budapest, Hungary, August 1996.

[SUG 01] SUGANUMA T., LEE S., KINOSHITA T., SHIRATORI N., "An Agent Architecture for Strategy-centric Adaptive QoS Control in Flexible Videoconference System", *New Generation Computing*, vol. 19, no. 2, p. 173-192, February 2001.

[TRZ 02] TRZEC K., HULIJENIC D., "Intelligent Agents for QoS Management", *Proceedings of AAMAS'02*, p. 1405-1412, Bologna, Italy, July 2002.

[TSA 00] TSATSOULIS C., SOH L., "Intelligent Agents in Telecommunication Networks", *Computational Intelligence in Telecommunications Networks*, p. 479-504, W. PEDRYCZ and A.V. VASILAKOS (ed.), CRC Press, 2000.

[URL 1] http://www.futsoft.com/pdf/fcapswp.pdf.

[URL 2] http://xenia.media.mit.edu/~nelson/research/routes.

[URL 3] http://www.tinac.com.

[URL 4] http://www.enib.fr/~harrouet.

[URL 5] http://www.j-sim.org.

[URL 6] http://www.ietf.org/rfc/rfc3086.txt.

[URL 7] http://www.fipa.org.

[URL 8] http://www.fipa.org/specs/fipa00037, FIPA Communicative Act Library Specification, Foundation for Intelligent Physical Agents, 2000.

Chapter 10

Agents and 3rd and 4th Mobile Generations

10.1. Introduction

Since the beginning of the 1990s, agent technology has become a promising field of research in several domains, particularly in intelligent networks. The impact of this technology in the telecommunications world must be analyzed. In particular, fast changes in the telecommunications market and the introduction of a new generation of mobile communication systems demand new technologies to respond to future user and service provider requirements.

In the last few years, agent technology has appeared in several fields: development of an intelligent user interface, personal assistance, distributed calculation, information research, telecommunication services and network service management. This new technology offers a promising solution to the complexity of distributed environments because agent solutions can:

– reduce traffic load (through autonomy and asynchronous operations carried out by agents);

– provide "on-demand" services adapted to user needs (by dynamic migration of the service provider agent towards the user terminal);

– increase flexibility, reuse and efficiency of software used to solve network problems;

Chapter written by Badr BENMAMMAR.

– personalize and integrate different services according to negotiated QoS;

– process a large volume of information.

In the next generation mobile networks, agent technology is expected to provide services such as capacity to maintain user preferences and profile information for each user.

The agent's role is to react dynamically and autonomously on behalf of the user in order to provide him with services and control applications.

Agent technology will probably reside over the Internet and software as well as chip cards, but it must also be able to act and cooperate with other forms of intelligence in the network to increase performances and accomplish more efficient operations.

This chapter is organized as follows:

– section 10.2 focuses on presenting agent technology;

– section 10.3 presents studies related to the applications of agents in UMTS;

– sections 10.4 and 10.5 address the application of agents in WLANs as well as 4th generation mobile networks.

10.2. Agent technology

10.2.1. *Definition of an agent*

In other works, we find several similar agent definitions. Their differences depend on the type of application for which the agent is designed. According to Ferber [FER 95], an agent is a physical or virtual entity operating in an environment, directly communicating with other agents, possessing its own resources, which is capable of partially perceiving its environment and has skills. According to resources, skills and communications, an agent will satisfy its objectives.

The authors of [CHA 99] have offered the following definition for an agent: an agent is a computer system located in an environment, acting in an autonomous and flexible way to reach its objectives.

In general, an agent represents a reusable software component providing controlled access to services and resources.

The behavior of each agent is restrained by policies which are defined by high level control agents.

Figure 10.1 represents an agent in its environment; the agent is activated by environment sensors when incoming and executes actions when outgoing.

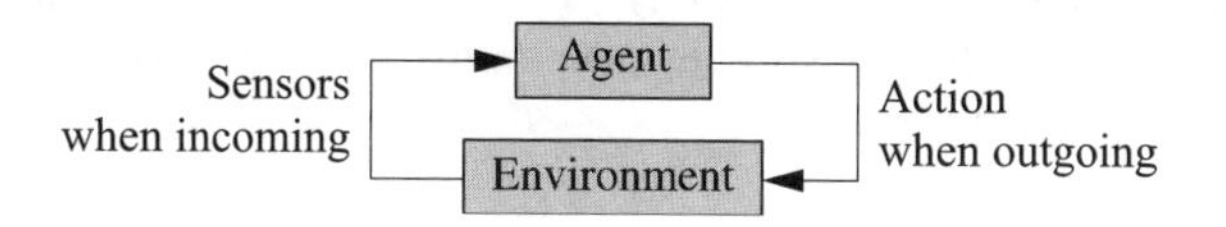

Figure 10.1. *The agent in its environment*

Jens *et al.* [JEN 98] have given the following definition: "an agent is a piece of software which is able to perform a specific predefined task autonomously (on behalf of a user or an application)".

Magedanz *et al.* identify two main types of agents [MAG 96]: mobile agents and intelligent agents.

10.2.1.1. *Mobile agents*

Mobile agents are software entities which can move within the network on their own initiative; they move from one computer to another and communicate with other agents, or they access server resources. Mobile agents have sparked great interest in the last few years for their capacity to support asynchronous interactions and to reduce network traffic during client/server interactions [BER 01].

In most cases, mobile agent systems provide a set of high level services such as monitoring functions, trip planning, resource management, introspection mechanisms and high level communication languages. Mobile agent technology also makes it possible to:

– provide asynchronous task execution. Because of this, dependence between clients and server applications can be reduced and automatic task processing is introduced;

– move a program towards a remote server for the development of a new type of distributed application. Services are no longer linked to a specific environment. Instead, they can be dynamically installed and used in the exact location where they are required;

– use mobile agent technology to help accomplish better resource development. By transferring applications from client to server and by executing local procedure calls instead of external calls, network traffic decreases.

The main attributes of a mobile agent are code, data and status (status of a process, computer or protocol) mobility. This enables software entities to move autonomously through the network to accomplish specific tasks, thus benefiting from proximity [JEN 98].

However, the main advantages of mobile agents are due to code mobility which makes it possible to accomplish asynchronous interactions and to decrease communication cost.

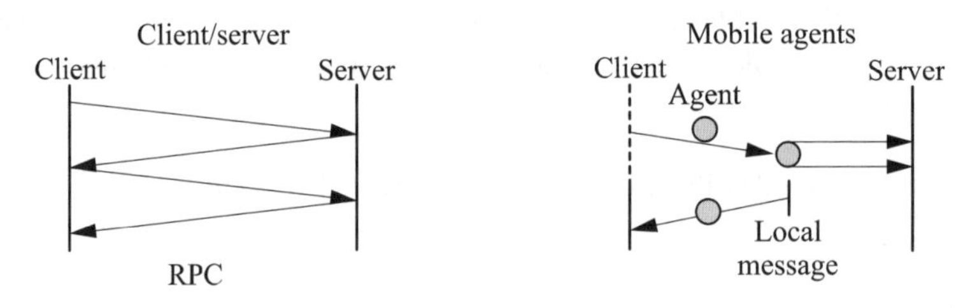

Figure 10.2. *Mobile agent paradigm*

Traditional distributed information technology is generally based on a client/server model which is often linked to the RPC (*remote procedure call*) mechanism.

When an RPC is called, parameters are sent to the server to execute the procedure and the result is then sent to the client. As described on the left side of Figure 10.2, each RPC requires the establishment of a communication channel.

The right side of Figure 10.2 shows the use of mobile agents; code mobility enables a client program portion to move towards the server, to act locally and come back to the client once the interactions are completed. Execution improvements depend on the network bandwidth and on the number of messages which can be optimized during transmission on the network.

A large number of applications can benefit from using mobile agents. Mobile agents can be a sophisticated solution for electronic commerce, database research, network management and mobile calculation.

Finally, security is always the weak link in mobile agent systems but is a promising research field.

10.2.1.2. *Intelligent agents*

An intelligent agent is a software entity which executes operations on behalf of a user (or another program) with a certain degree of freedom and autonomy and which, to achieve this, develops knowledge or representations of user desires and objectives [IBM 95].

Jens *et al.* [JEN 98] have given the following definition: "intelligent agents are software entities that are able to perform delegated tasks based on internal knowledge and reasoning, where aspects such as interagent communication and negotiation are fundamental". Generally, mobility is not considered as an intelligent agent property.

Another definition was given by Jennings *et al.* [JEN 98]: "an intelligent agent is a computer system, located in a specific environment, which is able to accomplish flexible and autonomous actions in order to respond to its design objectives".

A multi-agent system is an organized group of agents. It is made up of one or more organizations structured by cohabitation and collective work rules between agents. In a single system, an agent can belong to several organizations [RIV 98].

Existing multi-agent systems are made up of reactive or cognitive agents depending on the problem to process.

10.2.1.3. *Multidimensional characteristics of an agent*

An agent is characterized by [FER 95]:

– *nature*: an agent is a physical or virtual entity;

– *autonomy*: an agent is independent from the user and other agents;

– *environment*: it is the space in which the agent acts; it can be brought down to a network made up of all agents;

– *capacity of representation*: the agent can have a local vision of its environment but it can also have a larger representation of this environment and of the agents surrounding it;

– *communication*: the agent will generally be able to communicate with other agents;

– *reasoning*: the agent can be linked to an expert system or to other generally complex reasoning mechanisms;

– *anticipation*: the agent can generally have capacities to anticipate future events;

– *learning*: an agent will generally have a tendency to retrieve, store and reuse information extracted or received from its environment;

– *contribution*: the agent generally participates to problem resolution or to the system's global activity;

– *efficiency*: the agent must have quick execution and intervention capabilities.

10.3. Introduction to UMTS

The 3rd generation of mobile networks is a group of technologies developed to grow the 2nd generation cellular systems in terms of capacity, coverage, service personalization and quality of service.

1st and 2nd generation mobile networks are considered mobile phone networks, contrary to UMTS (universal mobile telecommunications system), i.e. a mobile network capable of offering multimedia services, everywhere and anytime. These services will be considered high quality and will be able to converge towards fixed, cellular and satellite networks.

UMTS relates to European efforts to establish a normalization of new mobile communications systems generation where personal services will be supported regardless of location, terminal, transmission method (wireless or wired) and network capacity. This is a unique and universal norm, able to offer global coverage and to ensure interconnection with fixed networks such as, for example, switched public telephone network (PTN), ISDN (integrated services digital network) or Internet. UMTS is expected to establish a simple integrated system in which users have flexible access to new sophisticated telecommunications services such as high throughput transmission for Internet/Intranet applications, electronic mail, video and e-commerce.

UMTS will provide user throughput of up to 2 Mbit/s and uniform access to services in all environments, such as residential, public and office, for example. The main characteristics of UMTS are illustrated in Figure 10.3.

UMTS users will be able to choose between massive service offerings from different networks and several service providers, and will have the possibility of using them from any computer and from anywhere, even while roaming.

Delivery of non-standard services through different types of networks presents difficult problems requiring innovative solutions. These problems have been identified by the European Telecommunication Standards Institute (ETSI) and International Telecommunications Union (ITU), and measures have been taken for roaming. The concept providing service portability is called *virtual home environment* (VHE). Its goal is to convert subscription services over different networks and the way in which they have been personalized. Service providers and administrators benefit from the flexible and simple introduction of new telecommunications services based on the VHE concept, which is one of the key points of the system. In this context, the agent technology represents a promising tool for network management offering new opportunities for the procurement and efficient deployment of telecommunication services.

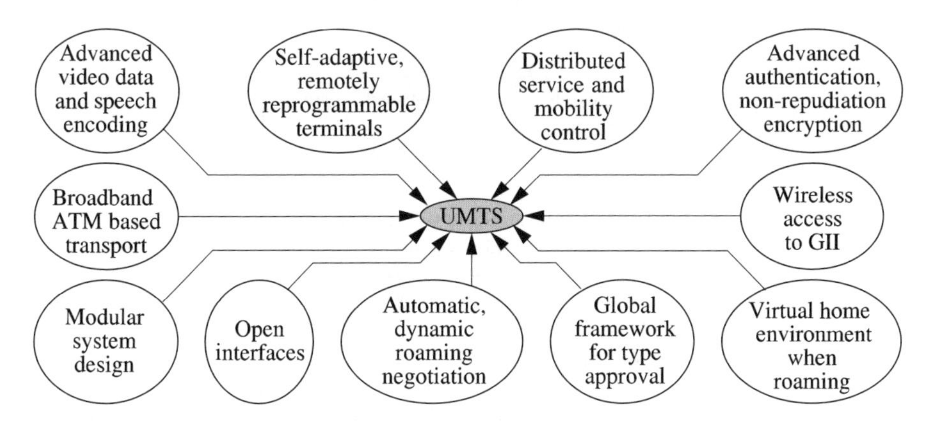

Figure 10.3. *UMTS characteristics*

10.3.1. *VHE*

GSM has had tremendous success [RIV 98] and among others we should mention roaming between the different networks and thus between countries using a unique registration.

This means that the subscriber is accessible by using a unique identifier and receives a unique invoice from his original service provider. The three most important requirements for future mobile users will therefore be identical to those of GSM users:

– easy handling of telecommunication services, including the possibility of subscribing to on demand services;

– global availability and performance compliant with telecommunications services;

– clear invoice with only one point of contact.

However, the future of telecommunications will not be homogenous and therefore these goals cannot be easily achieved. The goal of VHE is to resolve these issues by enabling a visited network to obtain information from the user's service provider during the registration procedure and any other information such as a personalized profile of the user service, which is necessary for services procurement. Although the physical achievement of a service can differ from one network to another, the VHE concept enables the user to access and use the service in the same way on any network (see Figure 10.4).

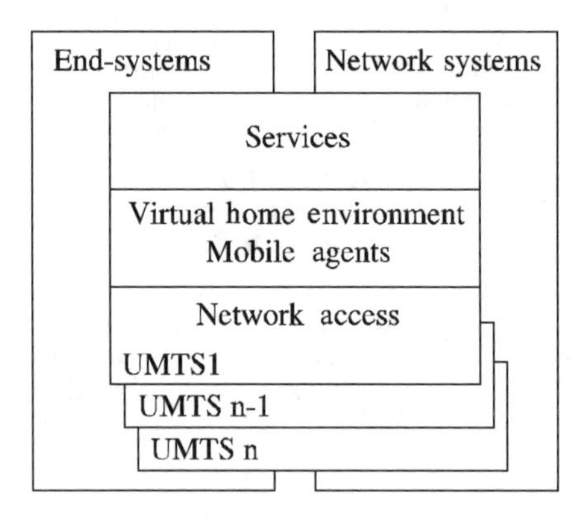

Figure 10.4. *Virtual home environment*

The VHE makes it possible for the user to be consistently presented with the same personalized service characteristics and user interface in any network, terminal and anywhere the user may be located. The capacity to provide this flexible service procurement has many implications for the network in terms of management, service control and signaling. GSM developments such as CAMEL (*customized applications for mobile networks enhanced logic*), MexE (*mobile station application execution environment*) and SAT (*SIM application toolkit*) respond to some VHE requirements and will also be adopted for UMTS. However, other techniques can be used to respond to all VHE requirements.

With agent technology, these advantages can be used in order to achieve the VHE concept.

10.3.2. *Application of agents in UMTS*

The 3rd generation mobile communication systems such as UMTS require new approaches to handle the large mass of data and to integrate the different existing communication technologies. Mobile agent technology is very attractive in developing a middleware layer for a distributed processing of mobile communication environments.

Mobile agents especially can play an important role:

– for an on-demand service procurement, UMTS presents the VHE concept so that users can roam and still be in their familiar environment regardless of their current location. It provides subscribed user services through different network domains and offers them various possibilities. This also includes support for quality of service (QoS) to adapt it to services in the current network. Tunneling of services in mobile agents offers a way to make services available; a user is usually expected to visit a network on demand and it adapts its services appropriately. In roaming, mobile agents can help a user negotiate a new QoS and adapt data flow to the new environment;

– for the network provider, mobile agents offer the possibility of reducing maintenance loads during service installation and operation. They provide decentralized development of service control and service management.

Mobile agent technology offers a software solution to the problem of integrating different core network technologies such as GSM and IP mobile.

Most studies focusing on the application of agent technology in UMTS attempt to define new services for accomplishing services procurement based on user needs, to improve different existing components in UMTS as well as to reduce bandwidth over radio interface for the different applications such as web access.

10.3.2.1. *Service procurement*

Agent technology is used for service procurement [JEN 98] based on VHE developed for the 3rd generation mobile communication UMTS system. In this context, two telecommunications services based on agent technology, i.e. APM (*adaptive profile manager*) and VAB (*virtual address book*), have been defined.

10.3.2.1.1. Adaptive profile management

Current applications in this domain only provide simple functionality for communication management, for example, forwarding all telephone calls to a predefined destination such as a voice mailbox or an alternative telephone number.

In future, by clicking on a simple key such as, for example, a do not disturb key, a user could automatically conduct his business by launching his APM to process communication requests received in a predetermined manner.

APM will provide the following possibilities:

– procurement of personalized management for user calls, i.e. a user can personalize call handling by creating a set of rules (*routing/filtering*) as a user profile element; this personalization is used by the system in order to intelligently manage communications;

– capability of combining remote services management with the management of all user profiles in a flexible manner;

– management of different terminals with many possibilities for profile management.

APM enables a user to control access to its communication services. It mainly depends on the following conditions:

– *time dependent conditions:* for example, a UMTS user who is in a meeting or has an appointment at the same time each week can place a time condition to ensure that all received calls are forwarded to his voice mailbox (for example, between 10 and 12 every Tuesday);

– *service dependent conditions:* in this case, the UMTS user can dynamically indicate how each type of communication received should be handled, for example, to reject particular fax messages, or to forward particular emails to another person depending on the sender and/or the subject of each email received;

– *user* (*screening*) *dependent conditions*: for example, a user can indicate a list of people who can reach him and have all other calls rejected or forwarded. The APM challenge is to dynamically adapt the presentation (GUI: graphical user interface) to the different computer screens. All user interface capabilities must be adapted to the computer capabilities.

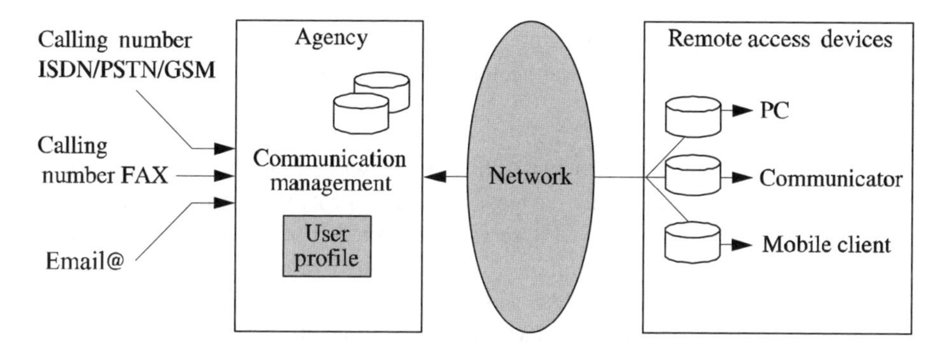

Figure 10.5. *Adaptive profile management*

In order to respond to the requirements mentioned above, an intelligent *communication agent* installed on the user's computer should organize and handle communication demands according to the user profile (see Figure 10.5). On the other hand, a *mobile agent* will be sent to computers to provide adapted GUI so that the user can remotely modify his profile from any point and any network, regardless of the platform used. The user will also be able to use this graphical user interface to be informed remotely of the last personal communication activities executed by his communication agent, for example, to view faxes or emails received and to listen to saved voice messages. The agent system can be developed in Java by using JTAPI (*Java telephony API*), JNI (*Java native interface*) and RMI (*remote method invocation*).

10.3.2.1.2. Virtual address book

The virtual address book (VAB) makes it possible for a UMTS user to use different types of computers to access and update his information. These computers do not have the same software and hardware characteristics and can be located in distant geographic areas.

Mobile phones usually have a personal memory in which alphanumeric names can be manually stored, but do not have the possibility of accessing new details in real-time. A desktop PC can provide access to the names of company employees and to their telephone numbers with the help of a specific application or more recently with an Intranet, but will require the user to launch a separate request to store a personal address book. Communication between these applications over the different equipment is not currently possible.

The agent architecture plays a very important role in providing a consistent view of user data regardless of the location, database, type of terminal and network configuration. The agent architecture also handles the problem of information uniformity as well as aspects relative to conflict resolution and shared access to address book information.

One of the main challenges with this service will be to provide an efficient interconnection between existing databases. Mobile agents represent a sophisticated solution. The architecture for VAB service is made up of multiple collaboration agents to provide a flexible solution for the problem of information uniformity, which is a key question for VAB service. The introduction of agents in this architecture provides additional value for the functionality offered by the storage technology that will be used to provide this service. This additional value is accomplished by the "intelligence" of agents processing user preferences, service characteristics, terminal and communication characteristics.

The main advantages in this regard are:

– a large distributed and heterogenous system such as UMTS uses a variety of storage technologies. The interconnection between these technologies is very difficult. Database agents eliminate this problem by providing an optimized solution achieved by collaboration between agents representing these technologies. For example, a user uses a file system to store information on his mobile computer and an object-oriented database on his PC at home. In order to obtain the updated information on the mobile computer, agents facilitate the negotiation with agents controlling information in the object-oriented database;

– an agent solution has a special advantage in wireless environments where bandwidth is a rare and expensive resource. For example, a mobile user would want to obtain an updated version of his information but that could be costly in a mobile network. With this architecture, it is possible to launch an update request and let the user agent assemble the information from different off-line hosts. In this scenario, if the user needs urgent information, the user agent can provide this information as soon as the user is online. For non-urgent updates, the user agent can provide the information when the fulfillment cost is within the limits indicated by the user.

In order to create a VHE in a visited location, the user agent can collaborate with agents in this new location to personalize this new environment according to user preferences. This requires interconnection between different storage mechanisms and presentation of environment information to create the VHE notion.

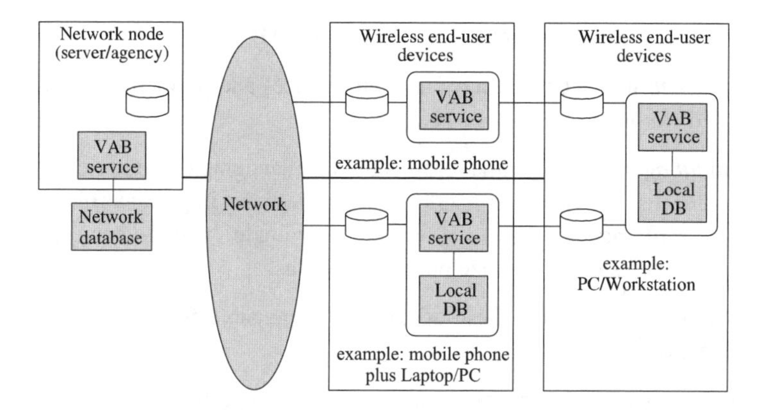

Figure 10.6. *Virtual address book*

Figure 10.7 shows a schematic agent architecture view. This architecture is based on a layered model in which agents representing service functionality are placed in the service layer. Each service is represented by a collaborating service profile agent by creating a VHE for this service in any environment where the service is supported. An example of agent task for VAB service is to maintain the host with which this service must synchronize in order to obtain updated user information. The information presenter agent considers user preferences, computer characteristics and service profile to show a consistent view of user data in an optimal mode. It is the responsibility of the information manager agent to communicate with the agent infrastructure layer to access and store user data.

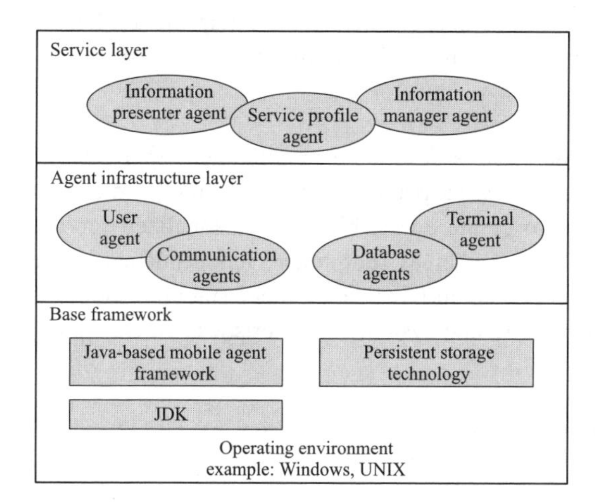

Figure 10.7. *Agent architecture*

At least four types of agents are necessary for VAB:

– *user agent*: this agent manages user preferences and plays an important role in creating VHE for the user;

– *terminal agent:* this agent is responsible for managing computer characteristics (for example, display type and size) and also plays an important role in creating VHE. *Terminal agent* and *user agent* coordinate with other service and infrastructure agents to personalize/optimize user tasks;

– *communication agents*: these agents are responsible for communicating with other applications;

– *database agents:* these agents are used to manage information storage. This type of agent provides information uniformity for VAB service. The storage mechanism used can change according to the type of terminal. This architecture reflects a peer-to-peer scenario. For a client-server configuration, the agent infrastructure layer will be divided and some components will be moved to the server side. Most of the database agents are moved to the server to handle multiple user requests. Server architecture must be equipped with additional capabilities to handle multiple requests from different clients.

10.3.2.2. *Improvement of different existing components*

In this case, the introduction of mobile agent technology in UMTS is mainly intended to improve existing network components. Since UMTS is a revolutionary concept, it should benefit from research activities and experiments done in GSM. In phase 2+ of GSM, CAMEL technology based on the IN (*intelligent network*) concept is introduced. Consequently, it is advisable to keep existing CAMEL functions, but to implement different interfaces corresponding to mobile agent technology.

In [PEY 99] the authors describe a mobile agent approach to achieve the VHE concept by considering two aspects, which are roaming and computer independence. Roaming attempts to create accessibility to personalized services in any network visited by the user, whereas computer independence focuses on adapting presentation of service to terminal equipment (mobile phone, laptop, PDA, etc.) according to its capacity and system characteristics, for example, and based on user profile.

For mobile agent technology integration in UMTS, the different existing components must be supported by additional interfaces.

Global UMTS architecture, integrating mobile agent technology, is illustrated in Figure 10.8. It contains network components such as MSC (*mobile switching center*) from the GSM architecture (MSC: element of core GSM network interconnecting control stations and enabling transmission of information in RTC by telephone circuit switching). However, MSC in UMTS will be more integrated than in GSM. It should control the different access networks, for example, existing GSM BSS (*base station subsystem*) as well as RNS (*radio network subsystem*).

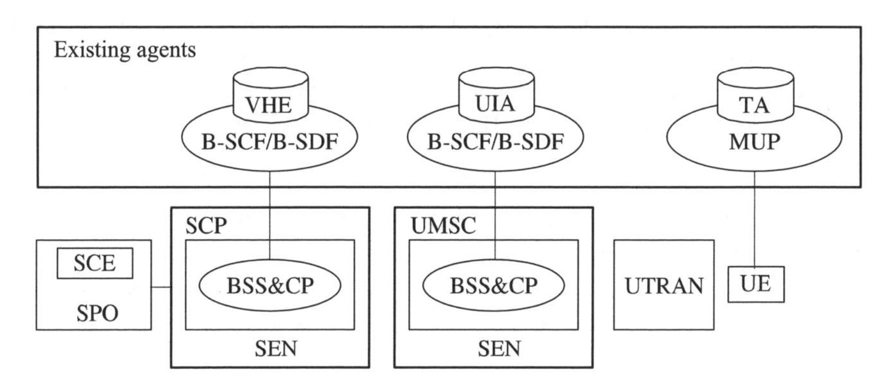

Figure 10.8. *UMTS agents*

For mobile agent technology integration into UMTS, SSP (*service switching point*) and UMSC SCP must be supported by other interfaces. SCF (*service control function*) and SDF (*service data function*) capabilities are merged to provide a distributed architecture. These combined SCFs and SDFs are called BSCF/B-SDF and provide interfaces for broadband communication services.

The agent execution environment should be provided by USIM or by TE/TA. In this architecture, the VHE agent provides personalized services for the user in any network and follows the user wherever he is.

The VHE agent contains a list of services subscribed to by the user as well as his profile. A *provider agent* (PA), residing in each environment, contains the knowledge of all services provided by this network. In this way, the VHE agent communicates with the *provider agent* during roaming in order to obtain information on new services provided by the visited network. The UMSC must provide an execution environment so that agents can accomplish their tasks. It makes it possible for a VHE agent, for example, to move between the different UMSCs. In addition, the VHE agent can be placed in the user terminal and move after establishing a connection in the new network.

The agent architecture contains other agents such as: TA (*terminal agent:* it provides an interface to the user for authentication before requesting a service), UIA (*user interface agent:* it transfers service characteristics to the terminal to provide appropriate GUI) and SA (*service agent:* it uses the service provided by the PA).

10.3.2.3. *Web access for UMTS*

Another application of UMTS mobile agents is web browsing which is among the most popular Internet applications and will certainly be among the first, together with email messaging, to furnish 3rd generation cellular networks [PER 02]. Web pages are hypertext documents containing other documents such as images, sound, texts and links to other web pages. HTTP is the protocol enabling access to web servers.

The wired/wireless convergence enables wireless administrators to offer new services already present in wired networks. However, this convergence between mobile communications and IP networks carries numerous problems because of the differences between these two types of networks in terms of capability and traffic [MOU 92].

Despite IP hardware independence, IP was not designed to function with mobile equipment. In fact, user mobility was not taken into consideration by the generation of IP protocols. Mobile IP was created for managing user macromobility and not micromobility. Besides, the narrow spectrum assigned to wireless systems and high error ratio on radio interface greatly reduce the available bandwidth and represent the main problem of connecting IP to a wireless environment. In a cellular context, mobile agents are used to improve web access and to decrease HTTP signaling on radio interface [PER 02].

The main characteristic of mobile agents is the ability to move from one site to another on their own initiative. Code mobility helps to reduce communications in a traditional client/server system in certain cases.

Requests sent by a user to obtain all the links in a web page will be slower during traffic overload. This delays web page display and increases user connection time in a radio interface. This task can be executed by a mobile agent in order to reduce occupancy in a radio channel.

Offline browsing consists of searching recursively for pages of a web site according to user filters. Mobile agents can act as an HTTP proxy in order to decrease the number of requests sent by a client and reduce the search delay for documents included in the same web page. In offline mode, mobile agents are used to search a website while the user is disconnected and provide the result once the process is completed. This should avoid long connection time and a large number of intermediate requests in radio interface.

Web pages are made up of HTML documents containing references to other pages and documents included as images, sound, video as well as other HTML files.

Offline browsers enable users with slow Internet access to completely or partially search for a web site based on filters. Users also upload each page as needed. In this way, they can access these pages in delayed mode once the complete transfer is finished.

Before presenting essential agent architecture in this context, in Figure 10.9 we present UMTS radio access network.

UMTS terrestrial radio access network (UTRAN) is made up of four fundamental elements (see Figure 10.9):

– user equipment representing the user's hardware. It consists of a mobile phone or a wireless laptop;

– cell which is the basic element of UMTS system cellular architecture. In fact, users can request services in all the cells created by the administrator in a geographical sector. In general, a cell is served by a base station;

– B-node made up of a set of adjacent cells;

– *radio network controller* (RNC) managing a set of B-nodes. RNC is responsible for handovers between all its cells. In a functional point of view in relation to the system, several identifiers are essential in this architecture: ID cell identification, RNC identification, RNC ID and *service area identifier* (SAI). SAI is necessary to calculate user location. User mobility is represented by two states: low activity and high activity. Depending on the activity of a UE, its location is known at cell level (in the case of low activity) or in a larger sector made up of several cells (in high activity). The service zone size is defined by the administrator. It is set according to the mobility management procedure used by the administrator. In [PER 02], a B-node is associated with the service zone.

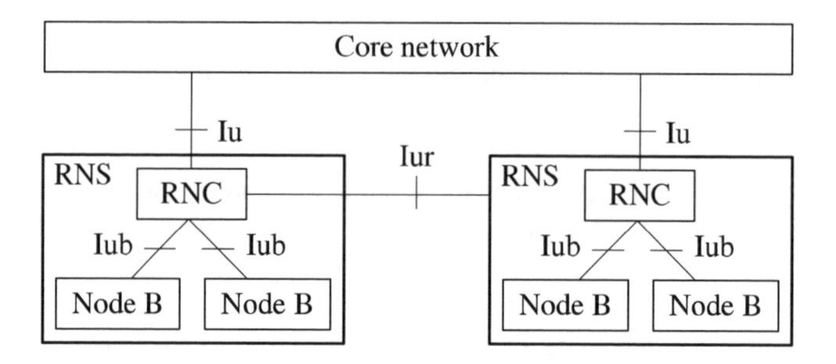

Figure 10.9. *UTRAN architecture*

In the UTRAN model, B-nodes are connected through the RNC element. The agents can use these connections to communicate and to move from one B-node to another adjacent B-node. Some wired UTRAN resources can be used during communication between agents. As illustrated in Figure 10.10, agents are installed in all B-nodes. Each agent will migrate from one B-node to another B-node by using special B-node connections. The choice of installing agents in B-nodes comes from the fact that the user updates his location each time he moves from one B-node to an adjacent B-node.

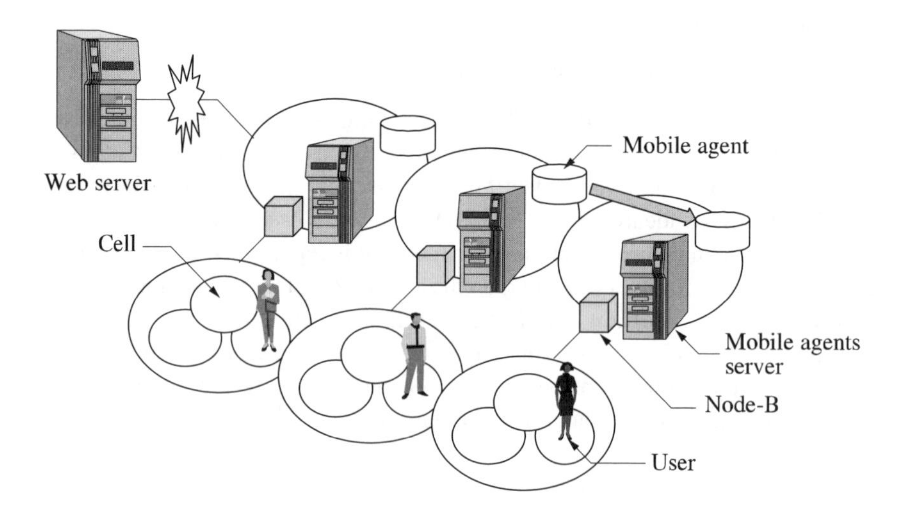

Figure 10.10. *Architecture with mobile agents*

A service zone location is equivalent to a B-node. In this context, the agent can use a path similar to that taken by the user. The architecture principle is based on the basic idea that a mobile agent is associated with each user in the cellular network. The agent carries out operations for the user based on a predefined profile. Traditional applications based on the client/server model require operations to be carried out on computer hardware. In this way, the user must remain connected during processing.

The use of mobile agent technology improves client/server interactions for the wired side of the network; radio connection is then only used if needed. This type of strategy decreases communication costs for the radio interface and improves communications by using the quickest connection for interactions with the server (bandwidth on radio connection is very low compared to bandwidth on a wired network).

The agent's task is to process the web page when it is transferred to the user. This process sends requests for documents included in the same page before the user requests them in a traditional way and the agent will send the response within a noticeably smaller delay.

The main problem with this technique is keeping the client from sending document requests that the agent has already sent. To solve this problem, [PER 02] use a small program on the terminal; this program processes the response received concerning documents included in the same page. When the agent finds a reference on a web page, it notifies the program of the incoming file.

In an offline browsing scenario, a user chooses the site or group of pages to download, then he places other parameters such as depth of recursion and file/url filters and sends his request to the agent. The agent will then download each page recursively and will analyze each page to follow the links. Since all the requests are executed by the agent, only the final result is sent back to the user in a simple file (.tar type file).

By using this technique, communications using the user's equipment only consist of accomplishing two tasks: processing user request and sending the final result; we eliminate all intermediate interactions executed by the agent. It is interesting to note that the radio interface user is not connected during the entire operation. Because of the tree structure of a web page, each page contains a group of documents as well as links to other pages. Consequently, a large number of requests can be ignored on the radio interface.

The following section illustrates the simulation result in [PER 02]. In both online and offline cases, the simulation model can be summarized in Figures 10.11 and 10.12, which correspond to a traditional and agent-based solution respectively.

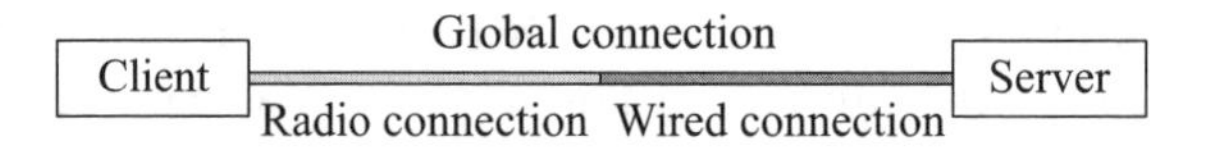

Figure 10.11. *Simulation model: traditional method*

Traditionally, the communication connection is modeled by a global connection made up of a radio connection and a wired connection.

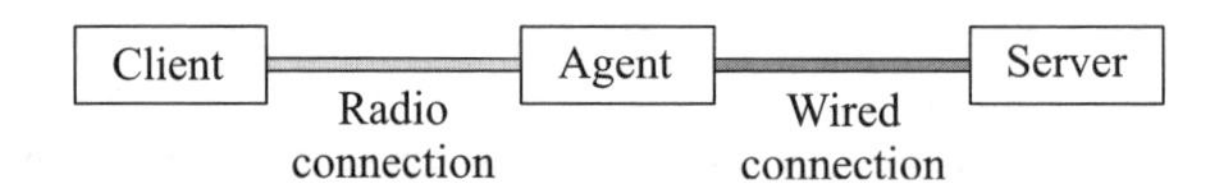

Figure 10.12. *Simulation model: using agents*

Taking agent mobility into account, the user can be seen as a static element in relation to this agent. The communication interface is modeled with different parameters such as transfer rate, propagation delay, buffer size at each node and the number of hops they contain. They calculate a transfer time for a specific volume of data based on these parameters. Three entities can be observed: client, agent and server. The server simulates a web server behavior, which provides the requested documents.

In the case of online browsing, in a traditional way and with no agents, the client simulates browser behavior, i.e. he calculates the time to send and receive responses from the server. In this model, the agent must simulate the same behavior on the wired connection and calculate the time for transmission requests and responses between user and server. In our case, radio interface communications are reduced at the demand of the main page and during receiving. Simulation is achieved to compare web access with and without agents. Different scenarios are considered to illustrate the usefulness of using mobile agents.

Table 10.1 represents the characteristics of processed web pages.

	Cache	Typical	Graphical	Text length
Average page size (kb)	9	6	6	30
Average link size (kb)	4	2	20	4
Average links per page	4	2	20	8

Table 10.1. *Characteristics of analyzed web pages*

In the case of offline browsing, the client calculates periods to request and receive pages and documents recursively. Documents included in the same page are downloaded in parallel with the main page, but complete pages are searched sequentially.

In conclusion and in [PER 02], the authors present a mobile agent architecture which makes it possible to decrease bandwidth and improve web browsing. The idea was to associate an agent with each user with this agent following each move of the user. Thus, a mobile agent architecture has been integrated to the UTRAN architecture.

Two scenarios have been studied in the HTTP context: online and offline browsing. Simulations prove that great interest can be expected with the use of mobile agents. In online browsing, the proposed architecture increases user bandwidth over radio interface and decreases page search time. In offline browsing, agents greatly decrease the number of messages in radio interface.

Finally, we can say that web browsing is an application which can benefit from the use of mobile agents. This technology also makes it possible to provide services adapted to user needs in 3rd generation mobile systems.

10.4. Introduction to WLAN

Wireless LANs (WLANs) are on the verge of becoming one of the main connection solutions for numerous companies. Large companies and several fields such as health and education are involved.

WLANs free the user from his dependence on backbone access, providing him with permanent and ubiquitous access. This freedom of movement offers several advantages to the user such as:

– ubiquitous access to the network;

– simple and real-time network access;

– faster and extended access to company databases.

WLANs also provide greater flexibility and enable implementation of transmissions in places where cabling is difficult, or even impossible.

However, this type of network does have its problems, we should mention:

– dynamic change in network and resource topology;

– complexity of mobility management;

– specific characteristics of radio channel;

– sudden and frequent disconnections;

– terminal resource limitation;

– highly variable delay and throughput since it depends on several factors such as number of users, interferences, multipath, etc.

10.4.1. *Application of agents in wireless networks*

To this day, there are very few studies on multi-agent systems in the wireless field. We can, however, mention the study on wireless mobile agent communications with some solution propositions [HEL 99], the use of intelligent agents for heterogenous network management [GUR 98], the proposition of a multi-agent architecture to control M-commerce [GRE 01], the definition of a reference scenario using fixed and mobile agents to guarantee QoS between a server and mobile equipment [ANS 00] and the implementation of a service of home banking operations based on the application of mobile agents [HAR 99].

More generally, five large fields of wireless agent technology use are identified.

10.4.1.1. *Localization of mobile hardware*

In a few years, wireless networks will provide multimedia applications with a large bandwidth with required quality of service. For this, a method will consist of using computer localization and make predictions to allocate resources necessary for terminals [ANS 00] and also to redirect traffic. The evaluation of mobile terminal localization becomes an integral part of wireless network management systems.

Several methods are proposed in other works to determine mobile terminal location in a wireless environment. They are based on AoA (*angle of arrival*), ToA (*time of arrival*), or RSS (*received signal strength*) measures [JAM 99].

A simple method of locating a mobile terminal is to use measures to determine radio path loss. In [MCG 03], mobile terminal localization is based on the propagation measure of the path loss between the mobile terminal and base stations (BS) by relying on RSS measures.

Agent technology can be used in this case to improve the results obtained and support terminal mobility, but in order to do this we must predict:

– future terminal localizations;

– QoS changes;

– disconnection times.

We must learn:

– the characteristics of network connection;

– the user behavior model;

– the user preferences.

10.4.1.2. *Efficiency improvement of a mobility protocol*

Agent technology can be used to improve the efficiency of a mobility protocol such as IP-mobile. This protocol was adopted by IETF to ensure mobility in an IP network. However, this protocol will not perfectly maintain the performance of a connection during user mobility. In fact, update database time indicating the new position of the user can deteriorate connection quality. Before the localization request arrives, several information packets can be lost if they are forwarded to the old mobile destination.

10.4.1.3. *Adaptation of handover to user requirement*

Agents can be used in wireless to support vertical handover. An agent, for example, that has deployed radio interface software for a technology can autonomously request (depending on user application needs) software for another radio technology before a vertical handover.

10.4.1.4. *Signaling control*

Mobile agent technology can also be useful for controlling signaling. The agents can also be used to dynamically negotiate user requirements of quality of service, security and mobility over the wireless network.

10.4.1.5. *Decrease of wireless access*

Mobile agents can distribute code to wireless network and mobile equipment. The number of exchanges needed in the network to provide a personalized service can therefore be reduced.

This enables the improvement of wireless performance by decreasing bandwidth consumption and by lowering delay time.

10.4.2. *Problems related to the application of MAS in wireless environements*

The main problems encountered during the application of agents in a wireless environment are related to security, cost, interoperability and implementation.

Security must in fact play an important role in the design of the environment supporting agent deployment because giving an important degree of intelligence and autonomy to an agent increases the risk of damages in case of function failure. In addition, in order to avoid any attack, access to internal network resources must be secured. A method of preventing security attacks is to authenticate user, agents and terminals. The cost in terms of agent migration must also be considered. Agent platform compatibility in terms of code and interfaces must also be assured.

Finally, implementing "mobile agent" technology in a wireless environment is a cumbersome task. However, the move is, in certain cases, necessary before the terminal on which it is located gets disconnected, for example.

10.5. 4th generation mobile network

10.5.1. *Definition of 4th generation*

The user of the 4th mobile generation (4G) uses several wireless access technologies. This user will want the possibility of being connected anywhere, anytime and with any access network. For this, the different wireless technologies, which are represented in Figure 10.13, must coexist in such a way that the best technology can be chosen based on user profile and each type of application and service requested by the user.

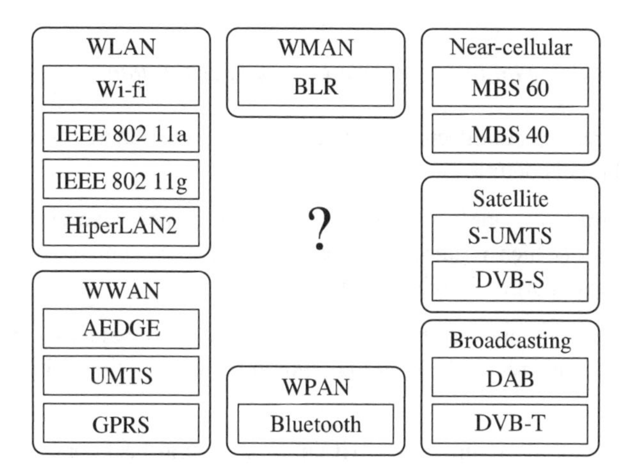

Figure 10.13. *Different wireless access technologies for the 4G user*

In this context, hardware equipment will need to permanently search for the best access network according to user needs. Agent technology can play a very important role in this choice.

10.5.2. *User expectations for mobile 4G networks*

For 4th generation mobile networks, a few potential scenarios have been identified and the common aspects are shown below:

– new input/output equipment will be available for quick data exchange;

– new semiconductor industry (4G terminals are available for everyone);

– access to 4th generation mobile systems at low cost;

– number of users will reach a high level;

– there will be strong competition between applications and service providers to satisfy users;

– Internet wired or wireless access quality will be equal or almost identical;

– multimedia applications will be used on a large scale;

– mobile networks should be stable, secure and available all the time.

Interconnection will need to be easy between the different systems (GPS, Internet, other communication networks).

10.5.3. *Technical conditions to achieve 4th mobile generation*

In order to respond to different user needs, 4th mobile generation must satisfy the following technical conditions:

– the majority of people can access voice or database services provided by mobile networks (which requires efficient resource management, for example, the use of an ad hoc extension in wireless systems);

– the mobile network can be completely linked to the Internet because of its basic concept (in this way, IP technology would be used by the mobile network (for example: VoIP));

– the network can organize itself (it controls several backbones and it uses the best);

– the system can maintain QoS (quality of service) parameters;

– the parameters of communication availability in the network must converge to 100%.

A hardware/software universal interface could be normalized, which should facilitate the development of new services without problems.

10.5.4. *Application of agents in 4G mobile networks*

Few studies are based on multi-agent systems in 4th generation mobile networks. Figure 10.14 gives an example of a network configuration providing several access technologies to the user. Agent technology is used to adapt horizontal (change within one access technology) and vertical (change of access technologies) handover to user QoS requirements [BEN 03].

The following example illustrates this principle.

EXAMPLE.– Our example is a wi-fi network deployed on a university campus where the agent technology is used to dynamically manage user mobility.

In a wi-fi environment, the access point strategy change requires four steps:

– discovery of a target access point;

– synchronization with the access point;

– dispatching of an authentication;

– establishment of association.

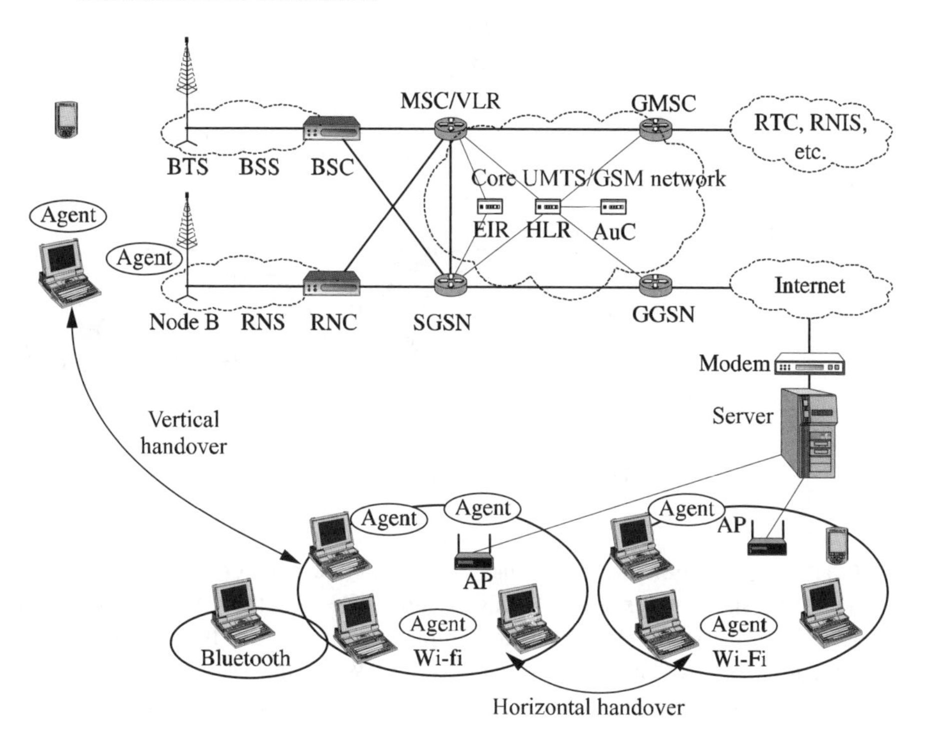

Figure 10.14. *Horizontal and vertical handover*

This strategy implemented in network equipment (BS and MH) is static, i.e. neither the service provider nor the client can change access point selection. However, in some cases this selection can be the wrong one. In Figure 10.15, mobile host 5 (MH5) on which a user launches a QoS heavy application (a video application, for example) receives the best BS2 signal. On the other hand, cell 2 is already loaded and consequently QoS required for MH5 cannot be ensured. A dynamic strategy consists of guiding MH5 to cell 1 which is empty and provide it with the required QoS.

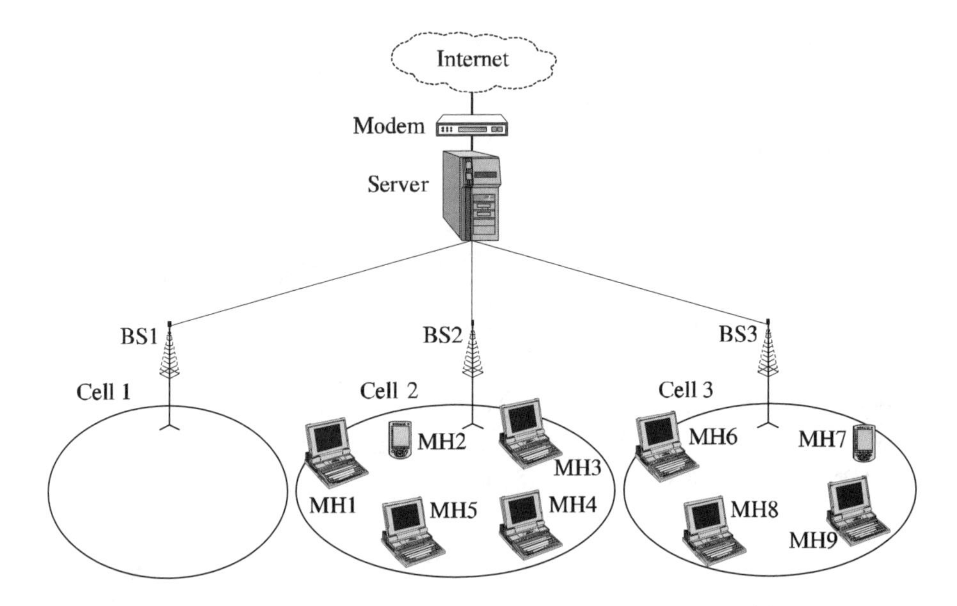

Figure 10.15. *Example of wi-fi network*

If all cells are full, the user must be able to use another access technology available to respond to his needs. In this example, UMTS is the technology that will be used, since the large number of users in place prevents wi-fi from responding to QoS requirements for the requested application. When the user launches another application that is not as critical in terms of QoS, or the wi-fi performance is acceptable for the application, the user must also be able to return to wi-fi technology (for example, because of high UMTS cost). The different vertical handovers must be totally transparent to the user and be based on application and user profile restrictions.

To provide QoS in a wi-fi connection, we must respect three principles [GAR 02]:

– the number of hosts authorized to use the channel must be limited;

– the geographical zone in which users communicate must be limited in order for all of them to use the highest throughput;

– the sources must be restricted by configuring traffic conditioners in the equipment.

In order to provide the required QoS to a multimedia application, we will respect these three principles and make three assumptions:

– from a specific number of users (N) grouped in one cell, the QoS required for a multimedia application will no longer be ensured and the cell will be considered as full;

– each access point contains a unique "location identifier". From this location identifier, a user can connect to the cell to ensure QoS required for the application;

– considering the work carried out by [MCG 03], an estimation of the MH position is done and an application distributing cells in each university room (conference room, library, etc.) is downloadable from the server.

Figure 10.16 illustrates cell distribution and associated identifiers in a conference room with three wireless access points.

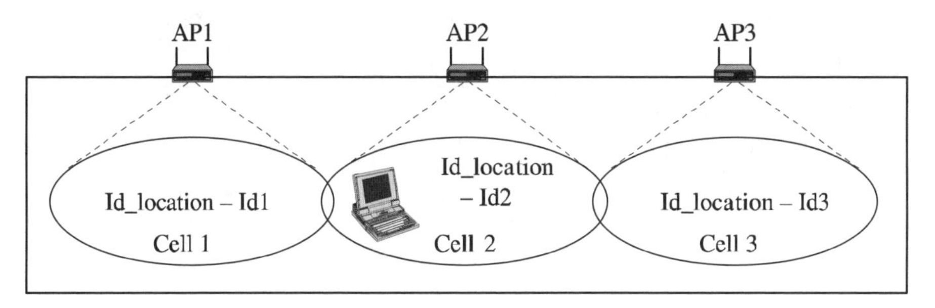

Figure 10.16. *Cell distribution in the room*

The multi-agent system contains two agents:

– *terminal agent*: this agent is located in the MH, it makes the connection between the user and system and it can be graphical or text. The terminal agent is autonomous; it activates during the launch of a multimedia application by the user and communicates with another agent on the access point in order to know the state of the cell, as well as that of the surrounding cells. It requests the deployment of another access technology if necessary;

– *status agent*: this agent is located in the access point and it determines the internal status of the cell as well as that of the surrounding cells. From N users grouped in the cell, its status will be considered as full. To know the status of surrounding cells, the status agent contacts the same agents in the surrounding cells and can then retrieve their states.

Figure 10.17 represents a group of system interactions. In this example, the user is located in cell number 2 of the conference room. He is, for example, viewing his emails, or making a file transfer. When launching a multimedia application, the terminal agent activates and sends a message (m1) to status agent in order to find out the status of the current cell. The status agent compares the number of users in the cell with number N and if the latter is lower than or equal to the number of users in the cell, it sends a message (m2) to the terminal agent to indicate that the current cell is full. At the same time, the status agent contacts the same agents in the surrounding access points (messages m3 and m4) to find out the status of neighboring cells. Each agent responds by a message containing the status of the cell or the number of users in the cell with its location identifier (messages m5 and m6). The status agent in the current cell makes a comparison between the number of users in the surrounding cells or between their states and whether there is at least one cell that is not full. It sends the location identifier of the chosen cell to the terminal agent (m7).

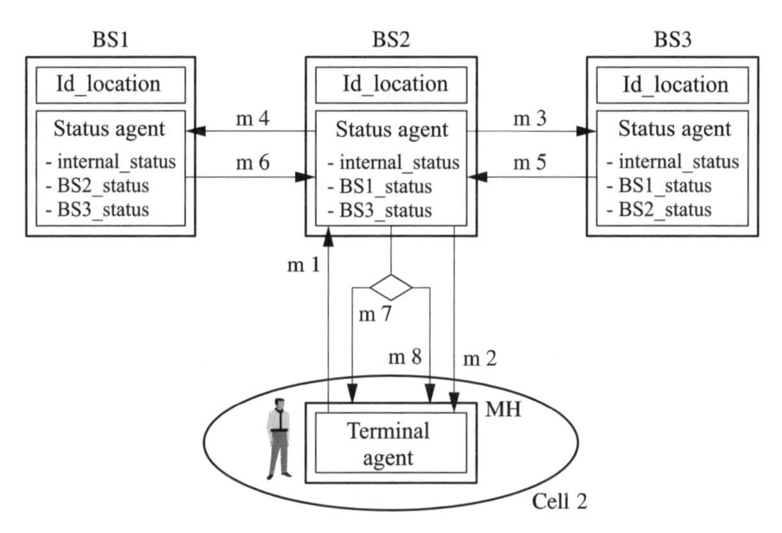

Figure 10.17. *Interactions between agents*

From this moment, the terminal agent sends a request to the server to download the application which will make it possible to find out where the cell involved is located in the room and notify the user.

On the other hand, if all the cells are full, the status agent contacts the terminal agent to communicate its situation (message m8) and from that moment, the terminal agent requests the deployment of UMTS.

10.6. Conclusion

Multi-agent systems represent a growing research theme. They involve several research fields such as artificial intelligence (AI), distributed systems, cognitive psychology and biology.

In this chapter, we have presented agent technology applied to 3rd and 4th generation mobile networks.

The application of agent technology in a 3rd generation mobile network such as UMTS attempts to define new services in order to provide services required by users, to improve different components within UMTS, as well as to reduce bandwidth on radio interface for the different applications such as web access.

Agents will mainly be used in wireless networks to improve existing mobility protocols and localization methods, to control network signaling, decrease access and adapt handovers to user needs.

In the 4th mobile generation, several wireless access technologies are available to the user. The latter wants to be connected anywhere, anytime and through any access network. In order to do this, the different wireless technologies must coexist so that the best technology can be used according to the user profile and each type of application and service required.

In this case, agent technology enables the user to change access networks according to his needs.

An agent can provide a service [JEN 98] based on each user profile. These agents often create new agents which move within the different networks to accomplish tasks defined by originating agents. It contacts service providers in order to determine the best offer or best application. Finally, they make a list of services found and appropriate. Agents can accomplish multiple tasks (data operation, auto-localization, locating other agents and users, etc.) because of programmed intelligence. Agents normally do not work on user hardware, they do not consume resources but they do operate on specific systems (Internet, for example); processed results may be easily displayed on a mobile computer screen.

Other agent applications in telecommunications are also important, such as the dynamic proposition or composition of personalized services for users.

10.7. Bibliography

[ANS 00] ANASTASI G., LA CORTE A., PULIAFITO A., TOMARCHIO O., "An Agent-based Approach for QoS Provisioning to Mobile Users in the Internet", *The 4th World Multiconference on Systemics*, Cybernetics and Informatics (SCI2000), Orlando, Florida, July 2000.

[BEN 03] BENMAMMAR B., KRIEF F., "La technologie agent et les réseaux sans fil", *Conference DNAC*, October 2003.

[BER 01] BERRY L., La société des agents, synthesis report, April 2001.

[CHA 99] CHAÏB-DRAA B., "Agents et systèmes multi-agents", Class, Information Technology Department, Laval University, 1999.

[FER 95] FERBER J., *Les systèmes multi-agents, vers une intelligence collective*, InterEditions, 1995.

[GAR 02] GARCIA-MACIAS J.A., ROUSSEAU F., BERGER-SABBATEL G., TOUMI L., DUDA A., "Différenciation des services sur les réseaux sans-fil 802.11", *Proc. Colloque francophone sur l'ingénierie des protocoles*, Montreal, Canada, 2002.

[GRE 01] GREGORY S., GUO Z., "The Architecture of Mobile Agent in Wireless Environment", *Macau IT Conference*, 200125-30, May 2002.

[GUR 98] GÜRER D., LAKSHMINARAYAN V., SASTRY A., "An Intelligent Agent-Based Architecture for the Management of Heterogeneous Networks", *DSOM'98*, Newark, USA, 1998.

[HAR 99] HARTMANN J., SONG W., "Agent technology for future mobile networks", *Second Annual UCSD Conference on Wireless Communications in Cooperation with the IEEE Communications Society*, San Diego, USA, March 1999.

[HEL 99] HELIN H., LAAMANEN H., RAATIKAINEN K., "Mobile Agent Communication in Wireless Networks", *Proceedings of the European Wireless'99/ITG'99*, p. 211-216, October 1999.

[JEN 98] JENS H., CARMELITA G., PEYMAN F., "Agent technology for the UMTS VHE concept", *ACM/IEEE MobiCom'98, Workshop on Wireless Mobile Multimedia*, Dallas, USA, October 1998.

[IBM 95] IBM, "Intelligent Agent", White paper, 1995, available at: http://activist.gpl.ibm.com:81/WhitePaper/ptc2.htm.

[JAM 99] JAMI I., ALI M., ORMONDROYD R.F., "Comparison of Methods of Locating and Tracking Cellular Mobiles", *IEE Colloquium on Novel Methods of Location and Tracking of Cellular Mobiles and Their System Applications 1999*, 1/1-1/6, 1999.

[JEN 98] JENNINGS N.R., SYCARA K., WOOLDRIDGE M.J., "A Roadmap of Agent Research and Development", *Autonomous Agents and Multi-Agent Systems*, vol. 1, no. 1, p. 7-38, 1998.

[KLU 97] KLUWE S., "Intelligent Communication Manager: conception and realisation of a platform-independent remote access over the Internet", PhD Thesis, Communication Networks, RWTH Aachen, November 1997.

[MAG 96] MAGEDANZ T., POPESCU-ZELETIN R., "Towards Intelligence on Demand on the Impacts of Intelligent Agents on IN", *4th International Conference on Intelligence in Networks*, Bordeaux, November 1996.

[McG 03] MCGUIRE M., PLATANIOTIS K.N., VENETSANOPOULOS A.N., "Estimating Position of Mobile Terminal from Path Loss Measurements with Survey Data", *Wireless Communications & Mobile Computing 3*, p. 51-62, February 2003.

[MOU 92] MOULY M., PAUTET M.B., *The GSM System for Mobile Communications*, 1992.

[PER 02] PERATO L., AL AGHA K., "Web Access for the UMTS Air Interface by using Mobile Agents", *IEEE WCNC'02: Wireless Communications and Networking Conference*, Orlando, USA, March 2002.

[PEY 99] PEYMAN F., CARMELITA G., FRANK B. "A Mobile Agent-based Approach for the UMTS/VHE Concept", *SMARTNET'99, 5th IFIP Conference on Intelligence in Networks*, Asian Institute of Technology, Thailand, November 22-26 1999.

[RIV 98] RIVADENEYRA J., MIGUEL-ALONSO J., "A Communication Architecture to Access Data Services through GSM", *7th IFIP/ICCC Conference on Information Networks and Data Communications*, Aveiro, Portugal, June 1998.

Chapter 11

Learning Techniques in a Mobile Network

11.1. Introduction

Because of the current evolution of society, people are increasingly on the move and need to communicate during their travels. This phenomenon has triggered greater demand and studies oriented towards the development of very sophisticated systems in order to respond to new user requirements. These requirements have indeed changed: if originally only voice was needed, wireless transmission demand providing reliable high definition sound, image and even high quality video communications has increasingly become popular with a large number of users. These users hope for mobility to be completely transparent in order to take advantage of performances similar to those from wired networks, despite the bandwidth greed of these new services.

Cellular systems are without a doubt those having experienced the strongest growth these last few years. The geographical zone served by a cellular network is divided into small surfaces called cells. Each of them is covered by a transmitter called "base station" (BS). Bandwidth in this type of network is divided into a separate group of radio channels defined by the access technique used. These channels can be used simultaneously as long as acceptable radio signal quality is maintained. This division can be done with different access techniques such as FDMA (*frequency-division multiple access*), TDMA (*time-division multiple access*), CDMA (*code-division multiple access*) or any combination of these methods [CAL 92, GIB 96, AGH01]. What is left now is to define the way in which these

Chapter written by Sidi-Mohammed SENOUCI.

channels are attributed to cells. There are two main methods: FCA (*fixed channel allocation*), where each cell has a specific number of channels and DCA (*dynamic channel assignment*) where all channels are grouped in a common pool (or group) and are dynamically assigned to cells [KAT 96]. In FCA or DCA type systems, a free channel not violating the constraint of channel reuse[1] is allocated to each user. However, when the user goes from one cell to another, he must request a new free channel in the destination cell. This event, called intercellular transfer or "handoff", must be transparent to the user. If the destination cell has no available channel, the call is disconnected. One of the great challenges for this type of network is managing this user mobility during a communication. In fact, the availability of radio resources for the duration of communication is not necessarily guaranteed and these users can experiment communication degradation or even break during intercellular transfer.

One of the major concerns during cellular network design is break probability decrease. In fact, from a user viewpoint, it is much more unpleasant than a connection failure. This is all the more important because cell size keeps decreasing in order to respond to cellular network growth, which considerably increases the number of intercellular transfers. Thus, since radio resource is a scarce resource, it is imperative to use it to the maximum, particularly in the case of a multiservice cellular network supporting several traffic classes where each one requires a different QoS level. For an administrator, it is always preferable to block a call from a lower priority service class (data, for example) and to accept another call with a higher priority service class (voice, for example). Consequently, a good call admission control (CAC) policy is certainly vital to maximize the usefulness of all these radio resources. To reach this objective, it is also necessary to find a good allocation method of all bandwidth available to all cells. These new mechanisms (CAC, dynamic resource allocation) must also handle frequent traffic conditions changes in cellular networks.

The objective of this chapter is to prove that it is possible to use techniques from the world of artificial intelligence (AI) and more specifically learning techniques in order to develop robust and efficient mechanisms to solve the problems encountered in cellular networks. These mechanisms must also exploit experience and knowledge which could be acquired during network operation.

In order to do this, we have developed new call admission control mechanisms in a cellular network considering both channel allocation diagrams: fixed (FCA) and dynamic (DCA). We have also developed a new dynamic resource allocation mechanism for choosing the best channel among all available channels in the

1 A channel can be used in many cells as long as the interference constraint is respected.

common pool, with the objective of maximizing the usage rate of all channels. These solutions are obtained by using the reinforcement learning Q-learning algorithm [WAT 89, WAT 92].

The rest of the chapter is organized as follows: section 11.2 briefly presents the notion of learning by emphasizing reinforcement learning and its application in telecommunications networks; section 11.3 presents a new call admission control method in cellular networks based on reinforcement learning; section 11.4 discusses a new dynamic radio resource allocation policy in cellular systems, also based on reinforcement learning; finally, section 11.5 concludes this chapter.

11.2. Learning

The term learning designates the capability to organize, develop and generalize knowledge for future use. It is the capability to take advantage of experience to improve problem resolution. Depending on the type of information available, two main approach categories can be observed. The first one, qualified as unsupervised learning, attempts to group objects into classes, relying on similarities. The second approach, i.e. supervised learning, is based on a learning group made up of objects where the class is already known.

11.2.1. *Unsupervised learning*

Unsupervised learning, also called learning from observation, consists of defining a classification from a group of objects or given situations. We use a mass of indistinct data and we wish to know if they have any group structure. The objective is to identify future data trends to be grouped into classes. This type of learning called clustering is found in automatic classification and in digital taxonomy. It searches for consistencies among a group of examples, without necessarily being guided by the use of acquired knowledge. It groups these examples in such a way that examples within one group are close enough and examples of different groups are different enough.

11.2.2. *Supervised learning*

In this type of learning, a teacher (or supervisor, hence the name supervised learning) provides either the action which should be executed, or an error gradient. In both cases, the teacher provides a controller with an indication of the action that it should generate in order to improve its performance. The use of this approach presupposes the existence of an expert able to provide a group of examples, called learning base, which is made up of correct associated situations and actions. These examples must be representative of the task to accomplish.

One of the variations of supervised learning, in which a "critique" of the calculated response is provided to the network, is called reinforcement learning (RL). This algorithm variation, explained below, has appeared as the most adapted to solve problems related to cellular networks and treated in this chapter.

11.2.3. *Reinforcement learning*

Reinforcement learning (also called semi-supervised learning) is a variation of supervised learning [SUT 98]. In contrast with the supervised approach, the teacher agent in reinforcement learning has a role of evaluation and not instruction. It is generally called critique. The role of critique is to provide a measure indicating whether the action generated is appropriate or not. The objective is to program an agent with the help of a penalty/reward evaluation without having to specify how the task must be accomplished. In this context, we must indicate to the system what goal to reach and the system must learn, by a series of trials and errors (in interaction with the environment), how to reach the set goal.

The components of reinforcement learning are the "apprentice" agent, its environment and the task to carry out (see Figure 11.1). The interaction between agent and environment is continuous. On the one hand, the agent's decision process chooses the actions based on situations perceived from its environment. On the other hand, these situations are influenced by these actions. Each time the agent accomplishes an action, it receives a reward. This reward is a scalar value indicating the consequence of the agent's action.

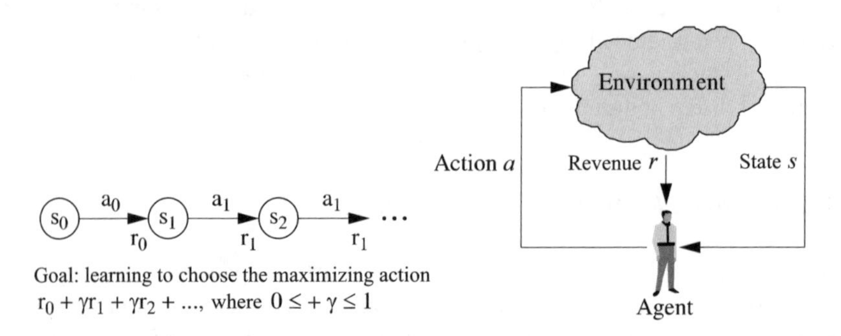

Figure 11.1. *Agent-environment interaction*

In a more formal way, we can designate s ($\in$ S, a finite group) as a representation of the actual environment state, a ($\in$ A, a finite group) as the action chosen and r ($\in$ R, a finite group) as the reward received. The interaction between agent and environment continuously consists of the following sequences:

– the agent observes the current state of environment $s_t \in S$;

– based on state s_t, the agent makes a decision by executing an action $a_t \in A$;

– the environment then makes a transition towards another state $s_{t+1} = s' \in A$ following probability $P_{ss'}(a)$;

– the agent instantly receives a specific revenue $r_t = r(s_t, a_t)$ indicating the consequence of this decision.

The agent's decision process is called policy and it is a function of the group of states to the group of actions (π: S $\rightarrow$ A). The agent must learn a policy π, which makes it possible to choose the next action $a_t = \pi(s_t)$ to execute, based on the current state s_t. The interaction between agent and environment is continuous and the apprentice agent modifies its policy based on its experience and its goal of maximizing the accumulation of rewards in time. This accumulation $V_\pi(s_t)$ achieved by following an arbitrary policy π, from an initial state s_t, is defined as follows:

$$V^{\pi}(s) = E\left\{ \sum_{t=0}^{\infty} \gamma^t r(s_t, \pi(s_t)) \middle| s_0 = s \right\} \qquad [11.1]$$

where E designates operator hope and factor $0 \leq \gamma \leq 1$ represents the temporal propagation constant.

For the agent, the objective is to maximize this sum of received reinforcements and its learning is done by several experiments. The agent is guided in this by different algorithms presented below.

11.2.3.1. *Resolution methods*

There are three fundamental method classifications for problem resolution of reinforcement learning: dynamic programming (DP), Monte Carlo (MC) methods and learning by temporal differences (TD) [COR]. Each class has its advantages and drawbacks. DP has well known/studied mathematical foundations but requires a complete and precise environment model. MC methods do not require models and are conceptually simple, but not adapted to an incremental step-by-step calculation. Finally, the TD approach combines the first two methods and uses in this way the best part of each one. This approach does not require a model and is incremental. These methods also distinguish themselves in terms of convergence speed and efficiency. Below we describe two methods of learning resolution by temporal differences: Q-learning and Sarsa.

11.2.3.1.1. Q-learning algorithm

Developed in 1989 by Watkins [WAT 92, WAT 89], Q-learning is part of reinforcement learning methods without model since the object is to learn by experimenting actions to accomplish according to the current state. The agent's goal is to learn a π: S $\rightarrow$ A policy which makes it possible to choose the next action $a_t = \pi(s_t)$ to accomplish based on current state s_t.

For each policy π, we associate a value $Q^{\pi}(s, a)$, that we will call its Q-value and represents the average expected gains if the action has been executed when the current system state is s and that π is then adopted as decision policy. Optimal policy $\pi^*(s)$ is the policy maximizing the accumulation of revenues $r_t = r(s_t, a_t)$ received after an infinite time. Q-learning algorithm's goal is to search for an approximation for $Q^*(s, a) = Q^{\pi^*}(s,a)$ in a recursive manner with only quadruple (s_t, a_t, s'_t, r_t) as available information. This information contains the state at moment t (s_t), state at moment t+1 ($s'_t = s_{t+1}$), the action taken when the system is at state s_t (a_t) and the revenue received at moment t (r_t) following the execution of this action.

Q-values are updated recursively at each transition by using the following formula [11.2]:

$$Q_{t+1}(s,a) = \begin{cases} Q_t(s,a) + \alpha_t \Delta Q_t(s,a), & \text{if } s = s_t \text{ and } a = a_t \\ Q_t(s,a), & \text{otherwise} \end{cases} \qquad [11.2]$$

where:

$$\Delta Q_t(s,a) = \left\{ r_t + \gamma \max_b \left[Q_t(s'_t, b) \right] \right\} - Q_t(s,a) \qquad [11.3]$$

from where we obtain the algorithm summed up in Figure 11.2.

```
Initialize Q0(s, a) to random values
Choose a starting point s0
While the policy is not good enough
choose at according to values Qt(st,.)
at = f(Qt(st,.))
obtain in return: st+1 (s') and rt
update Qt+1(st, at) by using formula [11.2]
End while
```

Figure 11.2. *Q-learning algorithm*

It is then necessary to set the ratio α in order to gradually establish the policy learned. Factor γ modulates the importance of the rewards expected. In [WAT 92], the authors demonstrate that if each pair (s,a) is infinitely visited and learning rate α leans toward zero, $Q_t(s,a)$ converges to $Q^*(s,a)$ with a probability of 1 when $t \rightarrow \infty$. The best policy will then be that with the highest Q-value:

$$\pi^*(s) = \arg \max_{a \in A(s)} Q^*(s,a)$$

The action choice (function f) is not described. It is possible to imagine different selection scenarios (exploration), for example, the action least used, or the one which returns the highest Q-value. In our studies, we have tested a group of exploration methods, such a $\in$-directed method, the Boltzmann method and even random strategies that we will discuss in more detail later.

11.2.3.1.2. Sarsa algorithm

Sarsa [SUT 98] is another algorithm for the resolution of a reinforcement learning problem. Contrary to Q-learning, during Q-values update the policy used for choosing the action at moment t is similar to that used for choosing action b at moment t + 1, or:

$$\Delta Q_t(s,a) = \left\{ r_t + \gamma\, Q_t(s'_t,b) \right\} - Q_t(s,a)$$

The exploration policy used can be $\in$-directed, for example. If each pair (s,*a*) is infinitely visited, Sarsa converges to the best policy.

11.2.3.2. *Application of reinforcement learning techniques – state of the art*

In what follows, we will list a few studies relative to traditional applications (robotics, games, etc.) and to some network applications (routing, CAC, etc.) using reinforcement learning as their solution. Unfortunately, there are not many studies relative to telecommunications network applications.

The first studies were achieved in the context of learning two-player games (checkers, backgammon, etc.). The other main reinforcement learning field is autonomous robotics, which is a scarcely addressed field in supervised learning because of the difficulty of modeling the real world with enough precision to account for the heterogenity of sensors, ambient noise and the robot-external world dynamic. In [SRI 00], the authors propose a multi-agent system (MAS) able to make economic decisions such as set prices in a competitive market context. They demonstrate that with Q-learning, the system is able to find the best price policy and to lessen the "price war" phenomena between different suppliers.

Studies [BOY 94, MAR 00, MAR 98] are routing propositions in telecommunications networks by using reinforcement learning techniques as a solution. The authors in [BOY 94] propose an extension of Bellman-Ford distance vector algorithm (BFDV) which they have called Q-routing. The reinforcement learning module is integrated into each node of a switched network. The routing policy attempts to find the best adjacent node to reach its destination with minimum "transmission time".

In [SEN 03c], we have proposed an ad hoc routing protocol Q-AOMDV based on one of the most important current ad hoc routing protocols, i.e. AODV (*ad hoc on demand distance vector*) [MANET]. The objective with Q-AOMDV is to balance energy consumption throughout the network by transmitting data traffic in different routes. The choice for the best route is achieved by using an adaptation of the Q-routing algorithm. Let us recall that AODV is an on demand ad hoc routing protocol, which only maintains one route towards destination. Contrary to standard AODV, results show that Q-AOMDV balances energy consumption over the whole ad hoc network while avoiding partitioning of the network into separate sub-networks.

A particularly interesting proposition for resource allocation has been proposed by Nie and Haykin [NIE 99]. We especially focus on this proposition because it is the subject of an extension addressed in section 11.4. The authors propose to solve the dynamic resource allocation problem in a GSM type cellular network (voice service only) by using Q-learning.

Finally, CAC problems in fixed networks and using reinforcement learning have been addressed in a few studies [MAR 00, MAR 98, MAR 97, RAM 96, MIT 98]. The authors of [MAR 00], for example, propose to solve this problem in a services integration network such as ATM (*asynchronous transfer mode*). The authors put themselves in the context where a service provider wants to sell network resources to maximize his revenues.

11.3. Call admission control

This section presents an approach which enables the resolution of the call admission control problem in FCA cellular networks supporting several traffic classes.

We recall that a geographical zone served by a mobile network is divided into cells sharing a frequency band in two methods: FCA and DRA [KAT 96]. In FCA, a fixed group of channels is allocated to each cell and a channel is allocated to each user. We also recall that one of the major concerns during mobile network design is the decrease of break probability during intercellular transfer.

Guard channel techniques decrease the probability of communication break by reserving channels for the exclusive use of handoffs in each cell [ZHA 89, KAT 96]. If these techniques are easy to size when we consider one traffic class (one telephone call), they become much more complicated to optimize and are less optimal in a multiclass traffic context.

In fact, in a multiclass traffic context, it is sometimes preferable to block a low priority class call and accept another call belonging to a higher priority class. Call admission control presented in this section enables this type of mechanism. It is obtained by using the reinforcement learning algorithm Q-learning which was presented in the previous section.

11.3.1. *Problem formulation*

We will focus on a simple FCA cell with N available channels and two[2] traffic classes C1 and C2. First class calls require one channel only, whereas second class calls require two channels. This cellular system can be considered as a discrete event system. The main events which can occur in a cell are incoming and outgoing calls. These events are modeled by stochastic variables with appropriate distributions. In particular, new incoming calls and handoffs are Poisson-ruled. Time spent on each call is exponentially distributed. Calls arrive in the cell and leave, either by executing a handoff to another cell, or by terminating normally. The network will then have to choose whether to accept or reject these connection requests. In return,

---

2 The idea can easily be extended to several traffic classes.

it retrieves revenues received from accepted clients (gains), as well as all revenues received from rejected clients (losses). The objective of the network administrator is to find a CAC policy which will maximize long-term gains and reduce handoff blocking probabilities and losses.

We have chosen the description of all states as $s = ((x_1, x_2), e)$, where x_i is the number of current class C_i calls and e represents a new incoming call or a handoff request in the cell. When an event occurs, the agent must choose one of the possible actions $A(s) = \{reject, accept\}$. At the end of a call, no measure need be taken.

For each type of call, revenue is associated with it. Since the main goal of the network administrator is to decrease handoff blocking probabilities, relatively high revenue values are assigned to them. Revenue values for class C1 calls are higher than those for class C2 calls, since C1 is presumed higher priority than C2.

The agent must determine a policy for call acceptance with the only knowledge being the state of the network. This system constitutes an SMDP with a finite set $S = \{s = (x, e)\}$ as space of states and a finite set $A = \{0,1\}$ as space of possible actions where Q-learning is the ideal solution.

11.3.2. *Implementation of algorithm*

After the call admission control problem formulation in the form of an SMDP, we will now describe two Q-learning algorithm implementations able to solve the problem. We have named them TRL-CAC (*table reinforcement learning CAC*) and NRL-CAC (*neural network reinforcement learning CAC*). TRL-CAC uses a simple table to represent function Q (the set of Q-values). On the other hand, NRL-CAC uses a network of multilayer neurons [MIT 97]. The differences between these two algorithms (TRL-CAC and NRL-CAC) are explained in terms of memory size and calculation complexity.

Since the approach uses a table, TRL-CAC is the simplest and most efficient method. This approach leads to an exact calculation and it is fully compliant with structures assumptions achieved in order to prove Q-learning algorithm convergence. However, when the group of state-action pairs (s, a) is large or when incoming variables (x, e) constituting state s are continuous variables, the use of a simple table becomes unacceptable because of the huge storage requirements. In this case, some approximation functions, such as state aggregation, neuron networks [MIT 97, MCC 43] or even regression trees [BRE 84] can be used efficiently. The neuron network used in NRL-CAC is made up of 4 entries, 10 hidden units and one output unit.

11.3.2.1. *Implementation*

When a call arrives (new call or handoff), the algorithm determines if quality of service is not violated by accepting this call (by simply verifying if there are enough channels available in the cell). If this quality of service is violated, the call is rejected; if not the action is chosen according to the formula:

$$a = \arg \max_{a \in A(s)} Q^*(s,a) \qquad [11.4]$$

where A(s) = {1 = accept, 0 = reject}.

In particular, [11.4] implies the following procedures: when a call arrives, acceptance Q-value and call rejection Q-value are determined from the table (TRL-CAC), or from the neuron network (NRL-CAC). If the reject has a higher value, the call is then rejected. Otherwise, the call is accepted.

In these two cases, and to find out optimal values Q*(s,a), the function is updated at each system transition of state s to state s'. For both algorithms, this is done in the following manner:

– TRL-CAC: [11.2] is the formula used to update the appropriate Q-value in the table;

– NRL-CAC: when a network of neurons is used to store function *Q*, a second learning procedure is required to find out neuron network weights. This procedure uses back propagation (BP) algorithm [MIT 97]. In this case, ΔQ defined by formula [11.3] is used as error signal which is back propagated in the different neuron network layers.

11.3.2.2. *Exploration*

For a correct and efficient execution of the Q-learning algorithm, all potentially significant pairs of state-action (s, a) must be explored. For this, during a long enough learning period, the action is not chosen from formula [11.4], but from the following formula [11.5] with a probability of exploration $\in$:

$$a = \arg \min_{a \in A(s)} \text{visits}(s,a) \qquad [11.5]$$

where visits(s, a) indicate the number of times configuration (s, a) has been visited. This heuristic, called $\in$-directed, significantly accelerates Q-value convergence.

These values are first calculated using this heuristic during a learning period. These values will then be used to initialize Q-values in both CAC algorithms.

11.3.3. *Experimental results*

In order to evaluate the advantages of our algorithms, we have used a discrete event simulation to represent the cellular network. Cell FCA has $N = 24$ channels. The temporal propagation constant γ has been set to 0.5 and exploration probability $\in$ to 1. Performance criteria used to compare these algorithms are: (i) gains, (ii) losses and (iii) break probabilities. Gains represent the sum of revenues from the acceptance of new calls or handoffs for both traffic classes in the cell. As for losses, they represent the sum of revenues from the rejection of new calls or cellular transfer failure. Break probabilities for all calls taken together (C1 and C2) have been calculated by using the following formula:

$$P_{HO} = \frac{\text{number of system handoff failures}}{\text{number of system handoff attempts}} \qquad [11.6]$$

As for break probabilities of traffic classes, they have been calculated for each traffic class C_i by using the following formula:

$$P_{HO(C_i)} = \frac{\text{number of system handoff failures for type } C_i}{\text{number of system handoff attempts for type } C_i} \qquad [11.7]$$

We have compared our policies to the one we called greedy policy [TON 00] (policy which accepts a new call or a handoff call if capacity constraint is not violated by accepting this call). We have also compared them to the guard channel mechanism. The guard channel mechanism enables the sharing of cell capacity between new calls and handoff calls by giving handoff calls higher priority. This is done by reserving, in each cell, channels for the exclusive use of handoffs (guard channels). The number of reserved channels for handoff calls is a very important parameter in this type of mechanism. In a multiservice context, the question is: how many channels can be reserved for handoffs for each of the traffic classes in order to maximize revenues? To answer this question, we have developed a traditional mathematical model where a cell is represented by a traditional multiserver M/M/N/N-type queue where each server represents a communication channel.

We have carried out a set of simulations including: (i) constant traffic load for all traffic classes, (ii) variable traffic classes and (iii) variable traffic load in time. We will not present all the results obtained, but for more information see [SEN 03a, SEN 03b]:

– constant traffic load: the first experiment considers a constant traffic load for both C1 and C2 classes. The simulation parameters used are the same as those used during the learning period;

– variable traffic load: in this second experiment we have used the same policies as the previous experiment (constant traffic load), but with six different traffic loads (for both C1 and C2 classes);

– variable traffic load in time: in this last experiment we have again used the same policies as the first experiment, but with a variable traffic load in time. Indeed, traffic load in a cellular system is variable in one day. We have used the traffic model presented in Figure 11.3 describing the variation of incoming rates during a working day. Rush hours appear at 11am and 4pm.

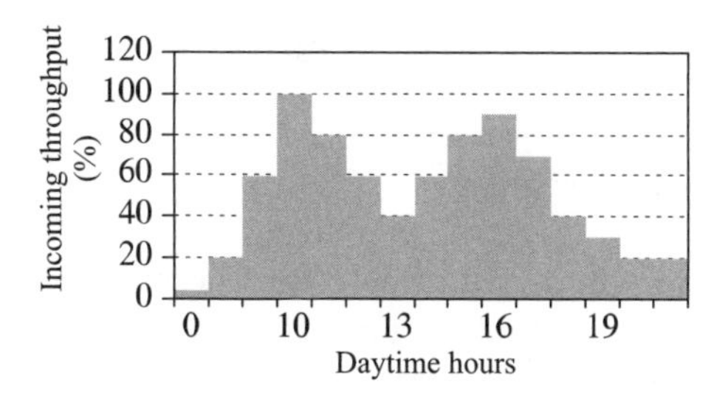

Figure 11.3. *Traffic model for a working day*

For all these experiments, the results have confirmed that the algorithms proposed are more efficient than the other heuristics for all traffic loads, in particular when traffic load is high. For example, Figure 11.3 shows simulation results with the assumption that both traffic classes used the same traffic model. Call blocking probabilities were calculated on an hourly basis. We have compared our algorithms to: guard channel with fixed thresholds (these thresholds have been calculated for a given constant traffic load) and guard channel with optimized thresholds (these thresholds have been calculated for each traffic load value). Improvements to proposed algorithms are apparent compared to greedy policy, particularly during traffic peaks (around 11am and 4pm). We notice that guard channel with optimized thresholds gives better results than the two algorithms based on Q-learning. On the other hand, this method presumes that optimized thresholds are calculated offline for each traffic load value. On the contrary, admission control algorithms based on Q-learning (TRL-CAC and NRL-CAC) have adaptation and generalization capabilities, so optimal Q-value values are not recalculated for each traffic load.

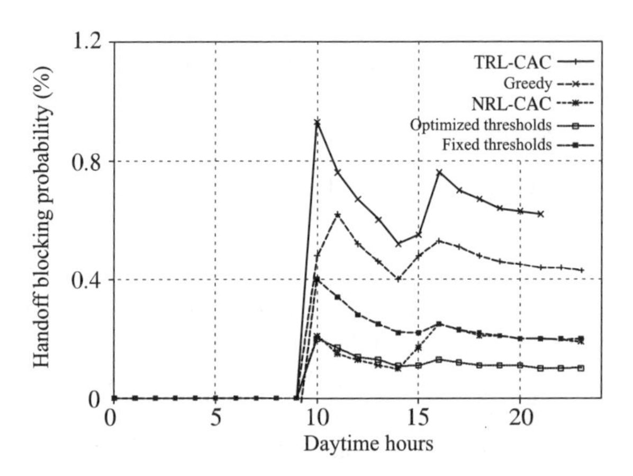

Figure 11.4. *Probability of handoff blocking with a variable traffic load in time*

11.4. Dynamic resource allocation

In this section, we will present a new mechanism enabling the resolution of dynamic resource allocation, also taking into account call admission control in DCA systems supporting several traffic classes.

Let us recall that in dynamic resource allocation (DRA) strategies, all available channels in the system are put in a common pool which can be used by all base stations [KAT 96]. During a communication request, a cell chooses a common pool[3] channel, which will be restored at the end of the call.

Different dynamic resource allocation algorithms have been compared in terms of performance, flexibility and complexity. One of the most widely used strategies in other works is a class of algorithms called "exhaustive searching DRA" [NIE 99, ZHA 89, COX 72, DEL 93, DIM 93, SIV 90]. In this type of algorithm, a reward (cost) is associated with each available channel. When a new call arrives, the system exhaustively searches for a channel with the highest reward (lowest cost) and assigns it to the call. Most of these propositions do not consider CAC policies as a method of preventing congestion [LEV 97]. The relation between resource

3 While respecting signal/interference relation *C/I*.

allocation and admission control has been previously studied [NAG 95, TAJ 88, YAN 94].

This section presents a new approach to solve the dynamic resource allocation problem, also considering call admission control in DRA systems. For reasons mentioned in the previous section, call admission control is vital when the network supports several classes of clients. Allocation policies are obtained by using the reinforcement learning algorithm Q-learning. The main functions of the proposed mechanism called Q-DRA presented in this section are: accepting clients and rejecting others, and allocating the best available channel for accepted clients. The goal is to maximize accumulation of received revenue through time.

We will now briefly describe the problem formulation by a SMDP and the implementation of the Q-learning algorithm capable of resolving this SMDP.

11.4.1. *Problem formulation*

This contribution is an extension of the study by Nie and Haykin [NIE 99] and is part of exhaustive searching DRA algorithms. We are considering the resolution of the dynamic resource allocation problem as well as the call admission control problem in a cellular network. This network contains N available cells and M available channels which are maintained in a common pool. It supports two traffic classes (contrary to [NIE 99] where only one traffic class – telephone call – was taken into account). Each channel can be temporarily allocated to any cell, as long as the constraint on reuse distance is satisfied (a given signal quality must be maintained).

The calls arrive in the cell and leave based on appropriate distributions. The network will then choose to accept or reject these connection requests. If the call is accepted, the system allocates one of the available channels to it from the common pool. The goal of the network administrator is to find a dynamic resource allocation policy capable of maximizing long-term gains and decrease probabilities of handoff blocking (contrary to [NIE 99] which does not give any priority to handoff calls).

We have chosen the group of states as being $\{s = (i, D(i), (x_1, x_2), e_i)\}$, where $D(i)$ represents the number of available channels in cell i where event e_i has occurred, x_k represents the number of current calls from class C_k and e_i indicates the arrival of a new call or a handoff call in a cell i. When an event occurs, the agent must choose one of the possible actions $A(s) = \{0 = \text{reject}\} \cup \{1, \ldots, M\}$. When a call ends, no measure has to be taken. The agent will have to determine the policies for accepting or rejecting a call and, when accepted, to allocate the channel that will enable the maximization of gain accumulation received by only knowing the current network state s. This system constitutes an SMDP which has as space of states a finite group $S = \{(i, D(i), x, e)\}$ and as space of possible actions a finite group $A = \{0, 1..., M\}$. The choice of revenues that we have used considers intercellular interferences and its consequence is a situation in which channels already used in compact cells[4] [ZHA 91] will have a better chance of being chosen. Contrary to other studies, Q-DRA considers the call type and grants priority to handoff calls.

11.4.2. *Algorithm implementation*

After the problem formulation in the form of SMDP, we will describe implementation of the Q-learning algorithm capable of its resolution. The cellular system studied is made up of N = 36 hexagonal cells and M = 70 channels available in a common pool. We use a reuse distance $D = \sqrt{21}R$ (R represents cell radius). This implies that if a channel is allocated to a cell i, then it cannot be reused in the two rows adjacent to i because of co-channel interferences. In this way, there are at most 18 cells interfering with each system cell. The temporal propagation constant γ has been set to 0.5 and the learning rate α to 0.1.

In the previous section, we have used a network of neurons to represent Q-values (NRL-CAC), but this time we have chosen an approximation based on state aggregation. Instead of precisely defining the number of calls x_i for each traffic class C_i, we have chosen to characterize traffic as follows: low, medium, high. The space of states is thus reduced and a simple table can be used to represent aggregated states. Because identical states (or states with a similar number of current calls) have the same Q-values, performance loss linked to aggregation becomes insignificant [TON 99].

11.4.2.1. *Implementation*

When a call arrives (new call or handoff call) in cell i, the algorithm determines if quality of service is not violated by accepting this call (by simply verifying if there are free channels in the common pool). If this quality of service is violated, the call is rejected, otherwise the action is chosen depending on the following expression:

$$a = \arg\max_{a \in A(s)} Q^*(s, a) \qquad [11.8]$$

where $A(s) = \{0 = \text{reject}, 1, 2..., M\}$.

---

4 Compact cells are cells with an average minimum distance between co-channel cells.

Formula [11.8] implies the following procedures. When there is a call connection attempt in cell i, Q-value of reject (a = 0) as well as acceptance Q-values (a = 1, 2…, M) are determined from the Q-value table. Acceptance Q-values include the different Q-values corresponding to choices of each channel a (a = 1, 2…, M) to serve the call. If the rejection has the highest value, then the call is rejected. Otherwise, if one of the acceptance values has the highest value, the call is accepted and channel a is allocated to it.

11.4.2.2. *Exploration*

For a correct and efficient execution of Q-DRA algorithm, the action is not chosen from formula [11.4], but based on a Boltzmann distribution [WAT 89] during a relatively long learning period. The idea is first to favor exploration (the probability of executing actions other than those with the highest Q-value) by using all possible actions with the same probability. Then the goal is to gradually move towards the use of the action with the highest Q-value. The values learned will be used later to initialize Q-values in Q-DRA.

11.4.3. *Experimental results*

In order to study Q-DRA performances, a group of simulations was completed. We have compared Q-DRA to greedy-DRA[5] policy [TON 99], as well as to the DRA-Nie algorithm [NIE 99]. Algorithm performances have also been evaluated in terms of gains, losses, as well as handoff blocking probabilities. A group of simulations was carried out including: a case of traffic load evenly distributed over all cells, a case of load not evenly distributed, a case of variable traffic load in time and a case of equipment failure. We will not present all the results obtained (see [SEN 03a, SEN 03c] for more information):

– even traffic distribution: the first experiment considers a constant traffic load in the 36 cells for both traffic classes. We have used policies learned during the learning period, but with five different traffic loads (for both C1 and C2 classes);

– uneven traffic distribution: in this second experiment, we have used policies learned during the learning period but traffic loads in this case are no longer evenly distributed over the 36 cells. The average traffic load considered was of 7.5 Erlangs;

– variable traffic load in time: in the third experiment, we wanted to test Q-DRA performances when traffic load changes in time;

5 Greedy-DAC: policy which randomly chooses a channel to serve a call with no measure of interference. Each M channel has the same probability to be chosen for serving the new call.

– equipment failure in a DRA system: in the last experiment, we have simulated equipment failure due to some channels becoming temporarily unavailable. At the beginning of the simulation there are 70 available channels in the system. However, between 10am and 3pm, we have temporarily suspended 0, 3, 5 and 7 channels.

For all these experiments, results show the possibilities that reinforcement learning offers in order to learn the best admission and dynamic resource allocation policy. Policy results using Q-DRA indicate significant improvements compared to other policies. These improvements are also slightly better than those from DRA-Nie. Q-DRA has an aptitude for generalizing and adapting to changes in traffic conditions. For example, Figure 11.9 shows the impact of channel failure over call break probabilities by using the Q-DRA algorithm. We notice that Q-DRA has a certain robustness against equipment failure situations and easily adapts, especially in the case where 3/5 channels have been suspended.

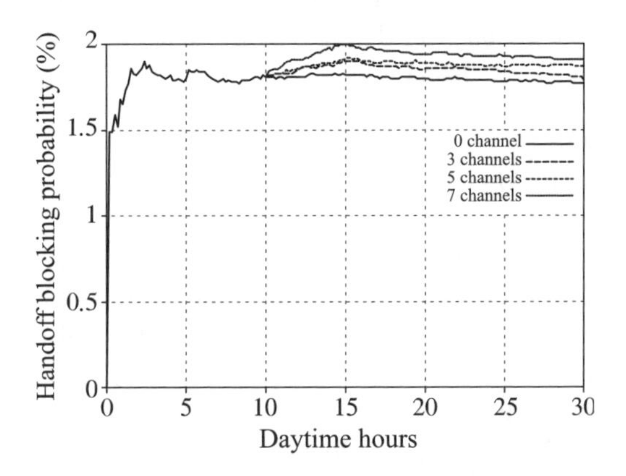

Figure 11.5. *Q-DRA performance during channel failure*

11.5. Conclusion

The first cellular networks were mainly designed to offer telephone service. Current cellular systems promise a diversification of services offered with clearly superior throughput. Service diversification (voice, SMS, multimedia services, Internet access, etc.) requires several levels of quality of service (QoS) to guarantee. However, availability of radio resources during a call is not necessarily guaranteed and mobile users can also experience service degradation/break. Since break/degradation of a current call belonging to a high priority service class is generally less desirable than break/failure of a call connection belonging to a lower priority class, new mechanisms of CAC are vital. In fact, efficient call admission control is required to prevent this limitation of available radio resources in the cellular network radio interface. Efficient management of radio resources through dynamic allocation policies also proves to be essential to prevent this type of problem. These new mechanisms consist of defining resource management rules for each traffic class, for the optimization of usage rate and to satisfy the multiple QoS constraints.

In order to prevent this type of problem, several propositions exist in other works and their main goal is to avoid inconveniences caused by communication breaks for users. However, we have noticed that these solutions often ignore experience and knowledge which could be acquired during real-time system execution.

The contributions presented in this chapter, relative to cellular networks, are intended to benefit from this experience and knowledge in order to optimize problems encountered in cellular networks. In the first contribution we needed to find a new approach to solve the CAC problem in a multiservice cellular network where channels are permanently allocated to cells. For the second, we needed a new approach to the dynamic resource allocation problem in a multiservice cellular network. This last contribution is ingenious, since it combines optimal CAC policy research and the best dynamic channel allocation strategy. These proposed mechanisms to solve such complex problems as those linked to cellular networks use reinforcement learning as their solution. These are creative and intelligent solutions. In addition to the creativity of these mechanisms, the advantages gained by using such approaches can be summarized as follows. Contrary to other studies (studies based on mathematical models or simulations, presuming fixed experimental parameters), these solutions are adaptable to variations of network state (i.e. variations of traffic conditions, equipment failure, etc.). Because of their distributed nature, they can easily be implemented in each base station, which makes them more attractive. Channel admission control and dynamic allocation tasks are quickly determined with little calculation efforts. They are obtained by a simple specification of preferences between service classes. We have also demonstrated, with a large group of experiments, that these mechanisms give the best results compared to other heuristics. These are distributed algorithms and signaling information exchanged between base stations are almost null. These mechanisms are therefore more attractive because of their implementation simplicity.

Finally, we can say that our main contribution has been to propose and test mechanisms for solving problems encountered in mobile networks (CAC and dynamic resource allocation). We have been able to demonstrate that it is possible to use techniques from AI and, more specifically, learning techniques in order to develop efficient, robust and easy to implement mechanisms.

11.6. Bibliography

[AGH 01] AL AGHA K., PUJOLLE G., VIVIER G., *Réseaux de mobiles et réseaux sans fil*, Eyrolles, 2001.

[BOY 94] BOYAN J.A., LITTMAN M.L., "Packet routing in dynamically changing networks: a reinforcement approach", *Advances in Neural Information Processing Systems (NIPS'94)*, vol. 6, p. 671-678, San Mateo, 1994.

[BRE 84] BREIMAN L., FRIEDMAN J.H., OLSEN R.A., STONE C.J., *Classification and Regression Trees*, Chapman & Hall, 1984.

[CAL 92] CALHOUN G., *Radio cellulaire numérique*, Tec & Doc, 1992.

[COR] CORSINI M.M., Class on reinforcement learning, available at: http://www.sm.u-bordeaux2.fr/~corsini/Cours/HeVeA/rl.html.

[COX 72] COX D.C., REUDINK D.O., "Dynamic channel assignment in two dimensional large mobile radio systems", *Bell Syst. Tech. J.*, vol. 51, p. 1611-1627, 1972.

[DEL 93] DEL RE E., FANTACCI R., RONGA L., "A dynamic channel allocation technique based on Hopfield neural networks", *IEEE Trans. Vehicular Technology*, vol. 45, p. 26-32, February 1996.

[DIM 93] DIMITRIJEVIC D.D., VUCETIC J., "Design and performance analysis of the algorithms for channel allocation in cellular networks", *IEEE Trans. Vehicular Technology*, vol. 42, p. 526-534, November 1993.

[GIB 96] GIBSON J.D., *The Telecommunications Handbook*, IEEE Press, 1996.

[KAT 96] KATZELA I., NAGHSHINEH M., "Channel assignment schemes for cellular mobile telecommunications systems", *IEEE Personal Communications Magazine*, June 1996.

[LEV 97] LEVINE D.A., AKYILDIZ I.F., NAGHSHINEH M., "A Resource Estimation and Call Adaptation Algorithm for Wireless Multimedia Networks Using the Shadow Cluster Concept", *IEEE/ACM Transactions on Networking*, vol. 5, no. 1, p. 1-12, February 1997.

[MANET] IETF MANET Working Group (*mobile ad hoc networks*), www.ietf.ora/html.charters/manet-charter.html.

[MAR 97] MARBACH P., TSITSIKLIS J.N., "A Neuro-Dynamic Approach to Admission Control in ATM Networks: The Single Link Case", *ICASSP'97*, 1997.

[MAR 98] MARBACH P., MIHATSCH O., SCHULTE M., TSITSIKLIS J.N., "Reinforcement learning for call admission control and routing in integrated service networks", in JORDAN M. *et al.*, (ed.), *Advances in NIPS 10*, MIT Press, 1998.

[MAR 00] MARBACH P., MIHATSCH O., TSITSIKILS J.N., "Call admission control and routing in integrated services networks using neuro-dynamic programming", *IEEE Journal on Selected Areas in Communications (JSAC'2000)*, vol. 18, no. 2, p. 197-208, February 2000.

[MCC 43] MCCULLOCH W.S., PITTS W., "A logical calculus of the ideas imminent in nervous activity", *Bulletin of Math. Biophysics*, vol. 5, 1943.

[MIT 97] MITCHELL T.M., *Machine Learning*, McGraw-Hill, 1997.

[MIT 98] MITRA D., REIMAN M.I., WANG J., "Robust dynamic admission control for unified cell and call QoS in statistical multiplexers", *IEEE Journal on Selected Areas in Communications (JSAC'1998)*, vol. 16, no. 5, p. 692-707, June 1998.

[NAG 95] NAGHSHINEH M., SCHWARTZ O., "Distributed call admission control in mobile/wireless networks", *PIMRS, Proceedings of Personal Indoor and Mobile Radio Communications*, 1995.

[NIE 99] NIE J., HAYKIN S., "A Q-Learning based dynamic channel assignment technique for mobile communication systems", *IEEE Transactions on Vehicular Technology*, vol. 48, no 5, September 1999.

[RAM 96] RAMJEE R., NAGARAJAN R., TOWSLEY D., "On Optimal Call Admission Control in Cellular Networks", *IEEE INFOCOM*, p. 43-50, San Francisco, March 1996.

[SEN 03a] S. SENOUCI, Application de techniques d'apprentissage dans les réseaux mobiles, PhD Thesis, Pierre and Marie Curie University, Paris, October 2003.

[SEN 03b] SENOUCI S., BEYLOT A.-L., PUJOLLE G., "Call Admission Control in Cellular Networks: A Reinforcement Learning Solution", *ACM/Wiley International Journal of Network Management*, vol. 14, no. 2, March-April 2003.

[SEN 03c] SENOUCI S., PUJOLLE G., "New Channel Assignments in Cellular Networks: A reinforcement Learning Solution", *Asian Journal of Information Technology (AJIT'2003)*, p. 135-149, vol. 2, no. 3, Grace Publications Network, July-September 2003.

[SIV 90] SIVARAJAN K.N., MCELIECE R.J., KETCHUM J.W., "Dynamic channel assignment in cellular radio", *Proc. IEEE 40th Vehicular Technology Conf.*, p. 631-637, May 1990.

[SRI 00] SRIDHARAN M., TESAURO G., "Multi-agent Q-learning and Regression Trees for Automated Pricing Decisions", *Proceedings of the Seventeenth International Conference on Machine Learning (ICML'00)*, Stanford, June-July, 2000.

[SUT 98] SUTTON R.S., BARTO G., ANDREW, *Reinforcement Learning: An Introduction*, MIT Press, 1998.

[TAJ 88] TAJIMA J., IMAMURA K., "A strategy for exible channel assignment in mobile communication systems", *IEEE Transaction on Vehicular Technology*, vol. 37, p. 92-103, May 1988.

[TON 99] TONG H., Adaptive Admission Control for Broadband Communications, PhD Thesis, University of Colorado, Boulder, Summer 1999.

[TON 00] TONG H., BROWN T.X., "Adaptive Call Admission Control under Quality of Service Constraint: A Reinforcement Learning Solution", *IEEE Journal on Selected Areas in Communications (JSAC'2000)*, vol. 18, no. 2, p. 209-221, February 2000.

[WAT 89] WATKINS C.J.C.H., Learning from delayed rewards, PhD Thesis, University of Cambridge, Psychology Department, 1989.

[WAT 92] WATKINS C.J.C.H., DAYAN P., "Q-learning", *Machine Learning*, vol. 8, p. 279-292, 1992.

[YAN 94] YANG W.B., GERANIOTIS E., "Admission policies for integrated voice and data traffic in CDMA packet radio networks", *IEEE Journal on Selected Areas in Communications*, vol. 12, p. 654-664, May 1994.

[ZHA 89] ZHANG M., YUM T.S., "Comparisons of channel assignment strategies in cellular mobile systems", *IEEE Trans. Vehicular Technology*, vol. 38, no. 1, p. 211-215, June 1989.

[ZHA 91] ZHANG M., YUM T.S., "The non-uniform compact pattern allocation algorithm for cellular mobile systems", *IEEE Trans. Vehicular Technology*, vol. 40, no. 2, p. 387-391, May 1991.

Chapter 12

An Experimental Example of Active Networks: The Amarrage Project

12.1. Introduction

The reference model for open systems interconnection from ISO has without a doubt been the starting point of the emergence of current networks. This layered model was warranted as its objective was to define a framework to unify different communication protocols and architectures. It has enabled a large community to have a unique reference point even when defining non-standard protocols. However, technology has since become more powerful, translating into large data volume transfer capabilities, among others. This has made it possible to consider the implementation of new applications, multimedia in particular.

Unfortunately, application requirements have become too diverse to consider a unique and generic process. The ideal would be to define a list of communication protocols specific to each class of application. *Active networks* are a first solution. In fact, they offer a new approach to network architecture, where network nodes/routers complete personalized processes on messages and therefore under application control. This approach has been motivated on the one hand by user-controlled applications where the processing of a network node is done by a network user and on the other hand by the appearance of mobile code and virtual computer technologies which enables network service innovations.

Chapter written by Nadjib ACHIR, Yacine GHAMRI-DOUDANE and Mauro FONSECA.

The *active network* concept was the result of considerations from DARPA (Defense Advanced Research Project Agency) research community in 1994 and 1995 on the future of networks. Several problems concerning current networks have thus been identified:

– difficulty of integrating new technologies and new standards in the network infrastructure;

– weak performance due to redundant operations at the level of several conventional layers;

– difficulty of integrating new services in the existing architecture model.

Because of this, many strategies represented as *active networkings* have now emerged to respond to these requirements.

Active networks constitute a program oriented network architecture approach in which messages transmit data as well as its executing code in the network. Network nodes are perceived as execution environments for instructions containing network messages. This approach causes the network to be called upon to handle applications and distribute tasks. This corresponds to a vision of operator network and business-to-business network, where the operator and user control the application's execution route. The active network approach is not actually new: we find these basic ideas in mobile agents and virtual machines. Such a network is a concrete accomplishment of these two approaches.

Currently, two additional active network approaches can be observed. The first is considered user oriented: each active network node is an execution context for a code determined by the user. This approach has been made a reality by the implementation of platform ANTS (*active network transport service*) [WET 98], which was developed and distributed by MIT. This implementation is widely deployed and constitutes a network of active nodes, i.e. Abone. The other approach is operator oriented: in this case, a node is a component which is dynamically configurable by the operator for new service support. Currently, a great expectation and exuberance exist about active networks. We note the committed support from the American defense agency DARPA expressed by the launch of a large number of projects in the US. The Japanese manufacturers Hitachi and NEC have also developed architectures based on active networks and have experimented with applications relying on this technology.

In this chapter, we will present contributions to this new domain by the French community. More specifically, we will focus on the Amarrage project [PRO 00] from the RNRT program supported by the Research Ministry in France. The objective of this project is to explore the potential of active networks for various problems in transport, quality of service, video and network management. Then, we will present an outcome resulting from the Amarrage project, which is an active DiffServ network control architecture. The goal of this architecture will be to show the huge potential that this technology can bring.

12.2. Description of the Amarrage project

12.2.1. *Objectives*

Without a doubt, active networks offer new perspectives in the design and deployment of new services. For an efficient industrial acceptance of this type of architecture over company networks or business-to-business, a full scale active architecture design and experimentation is vital. From these, large scale validations surrounding security, administration, architecture of network nodes and integration of existing technology can be achieved before their deployment and use. The goal of the Amarrage project is to characterize these requirements in terms of contribution for significant applications (multimedia, multicast, transmission, administration and security).

The sectors more specifically addressed with the Amarrage project are:

– nationally controlling the basic active network platform by supporting the deployment of different levels of services and protocols (particularly network, transport and application);

– innovative service experimentations and protocols on the national platform. The protocols involved are:

- protocols for reliable multicast and partial transmission,

- application protocols surrounding multimedia communications or more specifically active images (H.263, MPEG4),

- network management protocols;

– enhancement of basic platform with new generic operations which are useful to active protocols (for example, application adaptability, topology information);

– validation of this platform in a multisite environment (Nancy-INRIA LORIA, Paris-LIP6-ENST-L2TI, Toulouse-LAAS) (see Figure 12.1).

12.2.2. *Contributions*

The Amarrage project is divided into five sub-projects. Each sub-project handles a particular theme on which the applicability of active networks proves to be highly beneficial. *Sub-project 1* handles the architecture necessary for the development of an active open network. *Sub-project 2* explores the contribution of active networks in multicast communications. *Sub-project 3* concerns video applications and *sub-project 4* handles active network contribution to network management. Finally, *sub-project 5* is responsible for the implementation of a test platform the size of France. Below we will address the objectives of each sub-project in more detail.

12.2.2.1. *Sub-project 1*

The objective of *sub-project 1* is the *design and development of active network architecture.* The goal of this sub-project is to define, develop, deploy and test the appropriate active architecture responding to requirements from services and applications involved, such as video broadcasting, management, multicast and security applications. We should note that security is an integral part of this architecture of active nodes and will be perceived as a service. In fact, active networks are located in an open network context. This context requires the definition and implementation of security policies. The choice of a policy depends on application requirements as well as constraints caused by the dynamic network infrastructure. This architecture must be easily deployed nationally for large scale experiments. The development, deployment and experimentation of this architecture in a consequent platform make up the ultimate goal of this sub-project.

12.2.2.2. *Sub-project 2*

The objective of *sub-project 2* is to take advantage of active network technology in order to develop communication services, traditionally linked to the transport level, which are highly efficient and adapted to application requirements. The emphasis is put on two services: *reliable multicast* and *partial order and reliability multimedia transport.*

In fact, the transmission level is generally associated with end-to-end information transfer reliability. This notion of end-to-end is available in point-point, point-multicast, or even multicast-multicast. All applications do not have the same requirements in terms of reliability. Some will require a completely reliable transport, others will be satisfied with a best-effort transmission and others will demand partial reliability.

In the case where an application requires a reliable multicast and where the network multicast service does not provide this guarantee (which is the case with IP multicast), a transmission protocol can then be used to make end-to-end data transfer reliable. Among these applications, many are liable to involve a large number of players, from a few hundred to thousands. This scaling problem, added to the dynamic nature of group population and topology, cannot be processed by traditional error control mechanisms. Recent works have seen an impressive number of reliable end-to-end multicast propositions, but unfortunately none of them is completely satisfactory.

In addition, and still with regard to reliability at the level of end-to-end information forwarding, the partial contribution of a transmission protocol does not need to be proven anymore. In fact, it is accepted that TCP and UDP protocols are not adapted to the requirements of multimedia data flow and therefore different qualities of service will not be taken into consideration. Whereas TCP only provides a totally ordered and reliable service, UDP only provides a minimum service (without order or reliability), however, a multimedia flow may only require partial reliability. As an example, an image lost on a video flow of 25 images/s is not necessarily penalizing. If this kind of loss is acceptable in view of flow QoS, retransmission wait time is then avoided, which enables earlier data delivery.

Reliable multicast and partial multimedia order are part of communication services that several applications only want to be able to develop. Even if it is possible to develop them at edge system levels, it is often accepted that implementing them within the network layer, i.e. on the internal network nodes, generally leads to better functionality and performance. The traditionally adopted approach to integrate new functionalities to the network layer consists of deploying new protocols. However, modifying existing protocols is a long and risky process, first for acceptance and standardization reasons and then because it will most often be necessary to carry out a manual synchronized update of all equipment, or else preserve backwards compatibility. An alternative emerges with active networks for rapid deployment of new services. The approach in this sub-project consists of applying the active network concept to implement, deploy and optimize transmission services.

12.2.2.3. *Sub-project 3*

Transmission (including broadcasting) of video flows over variable throughput networks has unresolved problems slowing down commercial services. Among these problems, we should mention variation of throughput based on traffic, diversity of terminals with their specific service and quality wishes, real-time adaptation (regulation, personalized services, etc.). The objective of *sub-project 3* is to explore new functions that video applications can use in an active network context. This study has been illustrated through two applications: videoconference (*H.323/H.263* [REJ 96]) and *MPEG4* video [MPE 98].

With videotelephony, the Amarrage project explores interaction possibilities between the network and the application, dynamically giving the encoder high volume of information on the state of the channel, in order to obtain a better image quality, which is more robust in relation to network risks. The study themes are:

– *measure of network state and interaction with encoder*: the idea is to provide several measures of the connection state used by the video flow, in order to inform the encoder as soon as possible so that it can react accordingly. This mechanism must provide a richer and faster information return than the one currently ensured by RTCP. It also enables a higher tolerance to connection quality fluctuation (throughput availability, etc.);

– *minimization of transient disruption*: this starts by an identification of video flow structure (to GOB level) without using decoding, with the help of indications in capsules. This identification, achieved by active nodes on the flow route, enables a possible cover up of certain pieces of the flow in intermediate nodes in order to ensure a quick retransmission between nodes in case of loss/errors.

The command/reaction mechanism available from the active network makes the system (application + network) more dynamic. It is in this context that a study on encoder control has been carried out, in order to find a better dynamic encoder setting from an automatic point of view: convergence, stability, delay, etc. The network and application are at the origin of regulation commands with the help of information provided on transmission quality (global throughput, losses) and of interactive requests expected by the application itself. As for adjustment parameters, they will mainly be: quantification steps, coding modes, determination of image types and definition of movement vector strategy.

With MPEG4 format image distribution, we have explored the possibilities of dynamic deployment of differentiated services (according to user characteristics, for example, their location). The demonstration intended is a MPEG4 server defining a unique context (container) in which, for all terminals, different images (contents) from respective active nodes will display. Practically, it is the implementation:

– of an application that will be deployed in active nodes to achieve the generation of a video flow towards a (group of) addressee(s) controlled by the server. Video contents can be stored beforehand and the distribution order will be ensured by the active node following a local scenario;

– of a dynamic deployment mechanism of these applications and data.

12.2.2.4. *Sub-project 4*

The objective of this sub-project is the definition and implementation of a monitoring application for the complete platform. This application is built on active components. This means that it uses capabilities of the platform that it manages (for example, the transmission of active packets) to collect information useful to the operator of one or more nodes and to conduct its monitoring operations.

The studies carried out are located at two levels:

– at "*network*" level, management application provides at least a management functionality of active nodes constituting the platform and their interconnection. This configuration management depends on topological information and platform status information. It provides active node functions and group discovery network functions for connected nodes, active node interconnection, access to state of nodes and modification of this state. This configuration management functionality is made available to users by a graphical interface which enables them to control and act on this information. In addition, the applications have topological information which is useful to their deployment and/or to their operation, such as multicast;

– at "*service*" level: in this case, the application provides support for design, composition and deployment of new services by a service provider. In this way, it enables the connection between a physical node operator, designers of active services to deploy over nodes and users liable to use them.

12.2.2.5. *Sub-project 5*

An active backbone (A-Bone) exists at the international level grouping a set of active nodes and constituting a virtual network over the traditional Internet network. Initially, only one node in France (at Loria) is connected to the A-Bone.

The objective of *sub-project 5* is to develop a large scale experimental network made up of several active nodes distributed in the French territory (in particular Nancy, Toulouse and Paris). This network would test the full scale efficiency of added value network services which are made possible by active networks and especially in the context of two applications: network management and video.

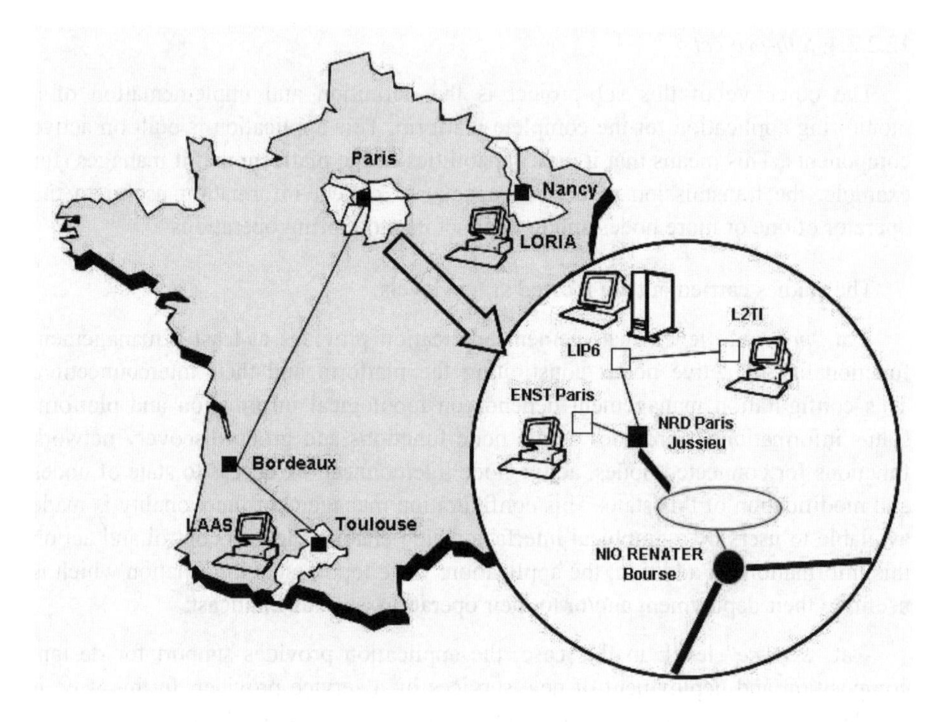

Figure 12.1. *Global Amarrage project platform*

12.3. Active networks: active architecture example for the control and management of DiffServ networks

Below we present one of the results obtained with the Amarrage project, i.e. an active architecture for the control and management of DiffServ networks (DACA: *DiffServ active control architecture*) (Figure 12.2).

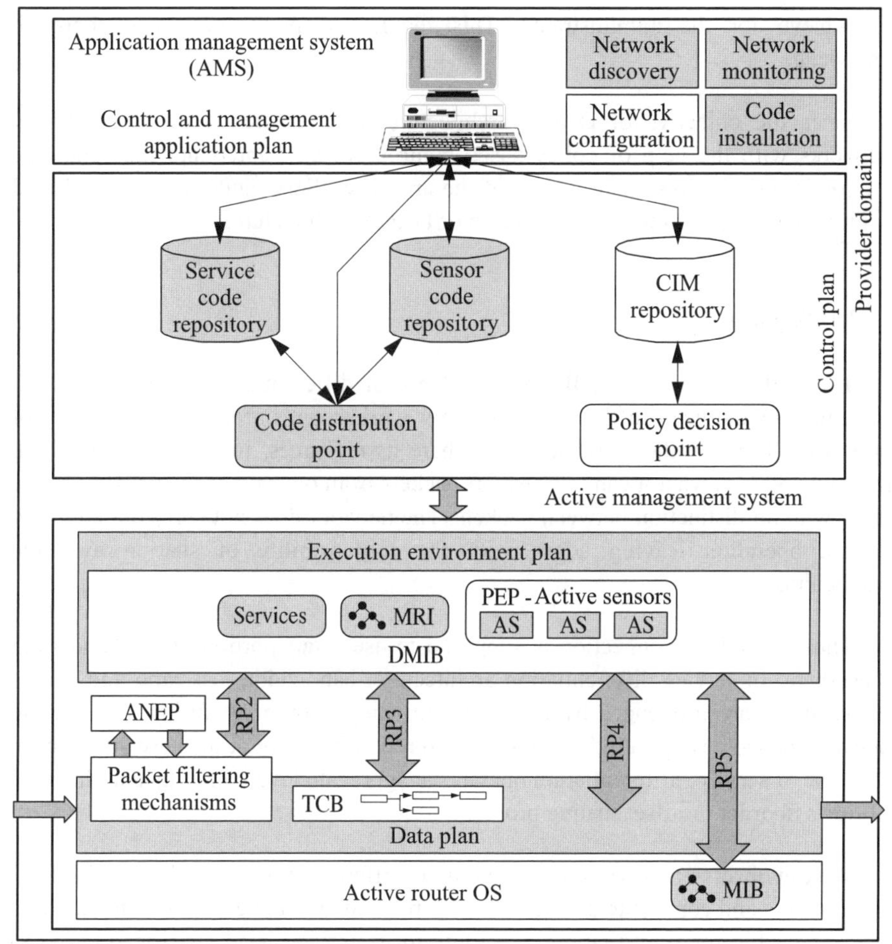

Figure 12.2. *DACA architecture*

This architecture has been proposed in the context of two sub-projects 1 and 4. The main objective is to facilitate and automate network equipment configuration through transparent deployment and dynamic services control in an active network. For this, we have focused more specifically on deployment and control of capabilities relative to the DiffServ model; we will then talk about the introduction of DiffServ service. Obviously, the goal is to reduce manual intervention overload for the maintenance and management of network nodes. In fact, in the current context of large heterogenous networks, this manual configuration is very complex to achieve for the administrator. This requires difficult work and thorough knowledge.

Since the objective of DACA architecture is to efficiently control DiffServ networks with the help of policy-based control, we will enter into the following section by briefly presenting a state of the art on DiffServ model and policy-based control, before going into the description of DACA architecture.

12.3.1. *DiffServ*

The DiffServ approach [BLA 98, NIC 99], or differentiated services, which is currently standardized in the IETF DiffServ workgroup, introduces a new way of processing flows in the network and to share its resources. In the current Internet, the network does what it can to transport packets from one end of the network to the other with no distinction between packets. The network does not carry out any flow control operation leaving edges with the responsibility of sharing available bandwidth.

Thus, each TCP connection is supposed to use a fair portion of the bandwidth they share. In service differentiation architecture, bandwidth, loss ratio and packet transit delay are influenced by traffic conditioning operations during input in the network, as well as by modifications done to router behavior at the network core. In this type of service, differentiation operates at aggregate level rather than at the level of flows in order to offset scaling problems.

We can observe two types of routers in the differentiation of service architecture: edge routers and core routers. Edge routers are located at the edges of a domain and are responsible for flow shaping according to its service level specification (SLS) and traffic classification by allocation of a label (DSCP: *DiffServ code point*) to each packet arriving in the domain. The value of this label for a given flow depends on its SLS and its instant behavior. Once labeled, the packet uses DSCP to choose the appropriate queue in core routers, as well as a suitable congestion process. The router behavior is therefore dependent on the DSCP.

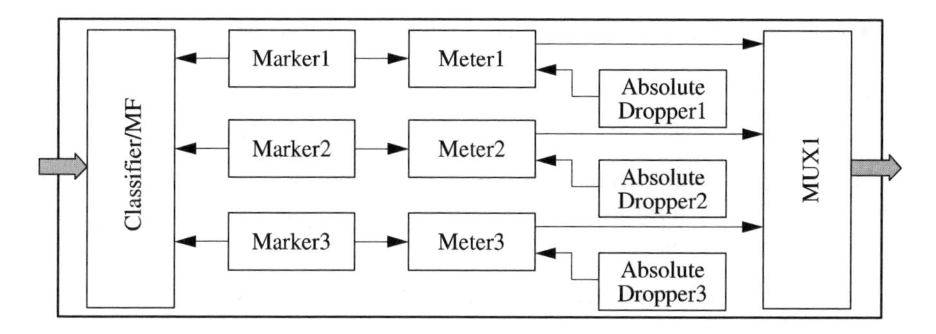

Figure 12.3. *Traffic conditioning blocks for an edge router*

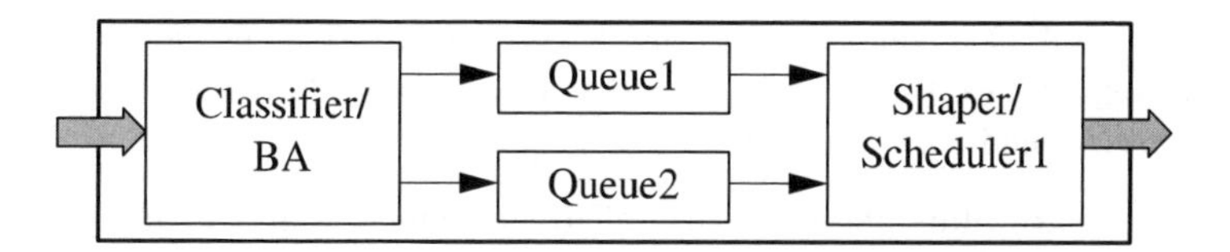

Figure 12.4. *Traffic conditioning block for a core router*

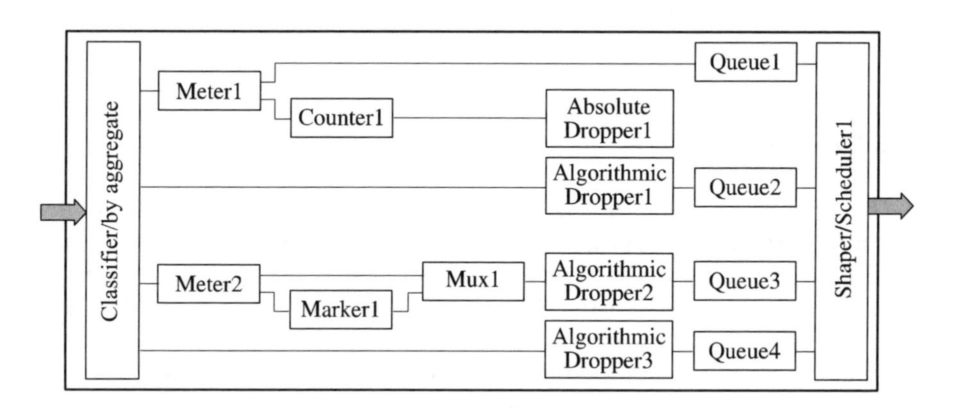

Figure 12.5. *Traffic conditioning blocks for the core router*

Figure 12.3 shows TCB (traffic conditioning blocks) for each type of router presented previously. Core router TCB can contain traffic shaping capabilities (Figure 12.4). These capabilities are necessary in the case where the network is not sized. They are similar to those of the edge router but they act on aggregates and not on flows.

12.3.1.1. *DiffServ service life cycle*

The introduction of DiffServ service is achievable as long as we are able to deploy TCB, which form the base of DiffServ service, in the different network nodes. In order to do this, we identify four phases for service configuration:

– *registration service phase:* this phase is initiated when a user signs a new SLA with an ISP to take advantage of a specific QoS. From this SLA, the network administrator creates corresponding policies and registers them in the *policy repository*;

– *pre-service phase:* in this phase, the network administrator must identify traffic conditioning components to be used. These components will then be connected together to form a TCB. This is done according to the service requested, existing services and network state;

– *service phase:* this phase consists of the introduction of a certain number of high level rules which enable an optimized network use. For example, if a congestion is reported in the network, then new TCB must be installed;

– *post-service phase:* in this last phase, the administrator focuses on client satisfaction and in particular on monitoring QoS that he offers him and detecting possible SLA violations.

12.3.2. *Policy-based control*

The term *policy* [RAJ 99] is largely used and it is therefore imperative that we give a clear definition of it in a network context. As a starting point, the term policy indicates a unified regulation for the access to resources and network services by relying on administrative criteria. Figure 12.6 shows different levels to which this type of regulation can be expressed or executed. The network view of a policy is explained in terms of end-to-end performance, connectivity and dynamic network states. This view is made up of several nodal outlooks to which correspond policy objectives and different node requirements. The policies themselves are made up of policy rules, which must be seen as atomic injunctions through which several network nodes are controlled. Finally, since each node has specific resource allocation mechanisms, each nodal outlook must be translated into specific instructions for a node's hardware devices.

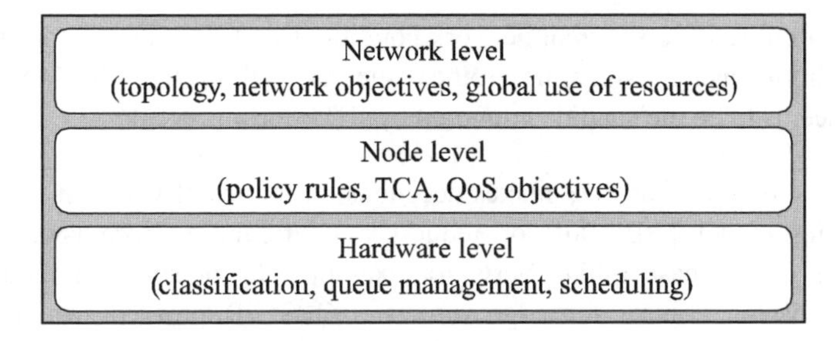

Figure 12.6. *Conceptual policy hierarchy*

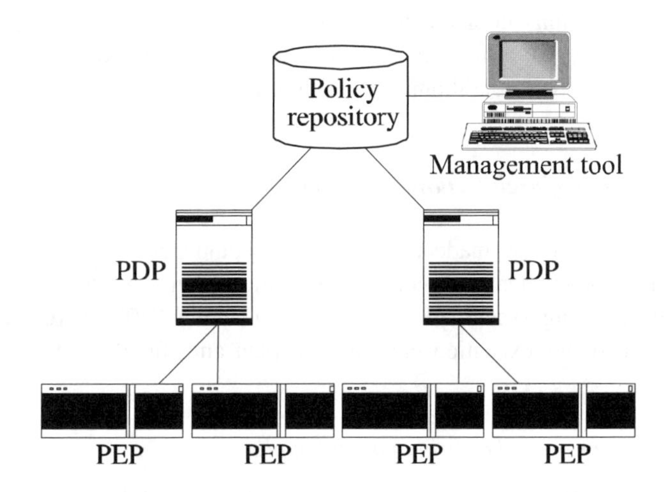

Figure 12.7. *Policy-based control architecture*

Figure 12.6 illustrates a generalized policy-based control architecture model. This figure shows the different components of a policy domain which will be discussed in more detail later.

A policy domain is a portion of the network managed by a set of policies. These policies are stored in an entity called policy repository. PEP[1] is the element responsible for the application and execution of policy actions and it is also the element which will control packet processing according to a given policy. The goal of PDP[2] is to determine which action must be applied to which packet. PDP interprets policy rules stored in the policy repository for one of more PEPs. In order

1 Policy enforcement point.
2 Policy decision point.

to do this, PDP relies on current network conditions and information transmitted by these conditions using a signaling protocol such as COPS [DUR 00]. In fact, COPS protocol seems to be the candidate favored by IETF for this task.

PEP may require that PDP makes decisions on its behalf when an event occurs (outsourcing model [HER 00]) or request a configuration (provisioning model [CHA 01]). In the case of DiffServ, for example, the outsourcing model would enable admission control for edge routers. COPS PR protocol would make it possible to transmit decisions to configure edge router and core router TCB. However, let us note that no mechanism exists today for the dynamic implementation of TCB statically implemented in routers and which use conditioning algorithms implemented in the hardware. Active network technology can therefore be used efficiently for the dynamic installation of TCB and to dynamically deploy new conditioning algorithms as needed.

12.3.3. *Description of architecture components*

DACA architecture is made up of a set of components distributed over five logical plans: management and control application plan, service provider control plan, active operating system plan (ADR-OS or active DiffServ router-operating system), active router execution environment plan and, finally, active router data plan.

12.3.3.1. *Management and control application plan*

In the *management and control application plan,* we find the *active management system* which is the central management point in the service provider domain. Its goal is to simplify DiffServ network control for the administrator. All management policies are defined and stored in this system. This system includes a specific number of tools (editing, distribution and management) in order for the network administrator to be able to define control and management capabilities to deploy in the network. These management capabilities involve both the network and services. Each of these capabilities can be deployed with the service composition principle, which means that the network administrator can specify high level capabilities developed around management components (information, service, sensor) pre-deployed in the active node. An incomplete list of these tools is: configuration tools to help the administrator configure the local active node as well as any other remote active node; network discovery tool is used to dynamically find active nodes and to decrease network topology; monitoring tool is used to monitor a particular active node or the whole network and finally code installation tool which enables the administrator to define new management capability and control algorithms. These are then deployed and installed in active nodes.

12.3.3.2. *Service provider control plan*

In the *service provider control plan*, we define the following components:

– *CIM repository (CIMR)*: represents the virtual repository for information relative to the network, services, user, SLAs, etc.;

– *services code repository (SCR)*: represents the virtual repository containing active codes relative to DiffServ TCB. Each code corresponds to an algorithm which can be dynamically installed in the active node. Each of these algorithms implements a specific traffic conditioning function (classification, marking, shaping, etc.);

– *sensor code repository (SeCR)*: represents the virtual repository of active codes relative to sensors. Sensors are management functions, which can dynamically be installed to accomplish a specific management function (monitoring, event notification, etc.);

– *policy decision point (PDP)*: is the element responsible for decision according to policies derived from network management strategy. It is designed to react to temporal events or the network by launching policy rules whose condition becomes true when these triggering events occur;

– *code distribution point (CDP)*: this component is responsible for the deployment of service and sensor codes. The request to deploy a new code can come from the network administrator or the PDP (policy decision).

12.3.3.3. *Active operating system plan*

The *active operating system plan* is an operating system which enables the node using it to support an execution environment. This plan also contains a conventional *management information base* (MIB). This base contains a group of objects representing the different active node resources which will then be perceived as an IP router.

12.3.3.4. *Active router execution environment plan*

The next plan, i.e. *active router execution environment plan,* includes the following components:

– *dynamic MIB* (*DMIB*), which is a specific active component for local on the fly installation and management of active management services, active sensors and MRI. Contrary to MIB, DMIB provides a view of the active router as an active node, i.e. management information and active services are perceived the same way. The active router can then be extended on the fly by new objects and their associated instrumentation;

– *managed resource information* (*MRI*), as with MIBs, is grouped in an information repository involving each resource in the active router. Since the MRI is passive, it requires instrumentation to update this information;

– *active sensors* (*AS*) push management information to MRI or to other management components defined by the administrator which can execute this instrumentation;

– *services* are active control services available within the active router and which can be used by the other components such as active sensors. Active services are also authorized to use information from managed resources, other services or active sensors. However, all services are not active all the time. A service must be activated to become operational. As an example, we can use the process for updating a FlowID (flow identifier) table. This table is used by the active router for filtering active packets as described in the following section;

– *active sensor PEP* executes the PEP tasks within the active router. Its role is to apply PDP and local PDP (LPDP) decisions. It has the responsibility to monitor MRI and locally apply policy rule actions when triggering events occur. These policies are defined in the form of rules with a condition part and action part. The LPDP then includes a specific number of active sensors. Each active sensor is responsible for the accomplishment of a rule and monitors the action part to verify if it is valid. It uses other components of the architecture (MRI, MIB, services) to monitor the active router. It also uses other active sensors to create a more complex monitoring process. An active sensor can be active or passive, depending on the state of its condition part. If it is verified, it will be active, otherwise it is passive in the sense that it will continue to monitor the validity of the condition part.

12.3.3.5. *Active router data plan*

Finally, in the *active router data plan*, we can find the packet filtering mechanism (PFM) and TCB. In order to optimize system performance and to avoid intercepting all packets in the network, we have defined a packet filtering mechanism capable of intercepting and redirecting packets according to specific filters. These filters define a set of active applications installed in the active node. In the data plan, installation, modification and withdrawal of a TCB are transparent (depending, of course, on system performance and its capacity to react to a specific event).

12.3.4. *Capsule filtering at the level of data plan*

When a capsule goes through the network, it goes through several active routers. If this capsule belongs to an active application whose protocol is deployed in a specific node, then it must be processed by the active protocol before being sent to the next node. We should also note that all active protocols are not installed in all active routers. Because of this, the role of PFM will then be to detect capsules which must be captured and to redirect them to the execution environment as fast as possible in order to avoid accumulation of delays. PFM uses a table of FlowIDs to store active protocol identifiers installed in the node. FlowIDs are transmitted either in option fields of IPv4 packets or in the field already planned for them (FlowID field) in IPv6 packets.

To maintain the FlowID table updated, the FlowID update service is activated. Its role is to monitor active protocols installed in the node and to update FlowIDs by adding or deleting entries in the table (Figure 12.8).

PFM plays a central role in the architecture and therefore has an impact in network performance. It must therefore operate as close to the active router's operating system as possible.

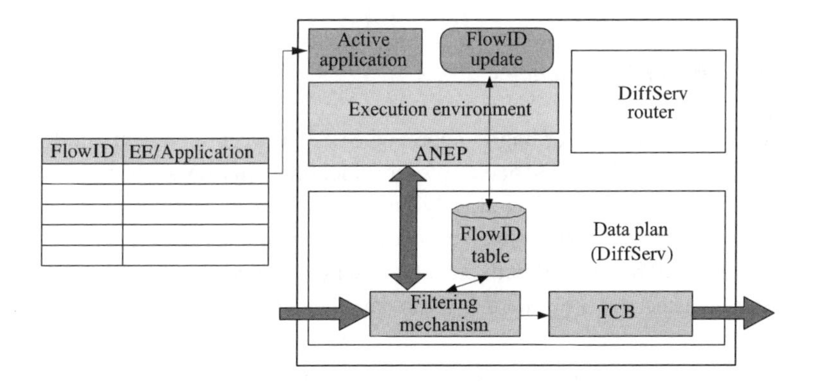

Figure 12.8. *FlowID table and PFM*

12.3.5. *Active router resource monitoring*

As previously established in this chapter, it is important that the network administrator has knowledge of network behavior. This knowledge can be perceived as the return or result of actions from the system administrator in the network. In the context of our architecture, DMIB is responsible for this task. In fact, it provides the set of capabilities required for dynamic monitoring and notification of events. We speak of dynamic notification and monitoring because, contrary to management in a traditional IP network with SNMP[3], the management information model is not static. DMIB enables efficient monitoring of any active router resource by using two expandable APIs which it defines as:

– *dynamic interface of information (DInfo):* Dinfo is an expandable object oriented API enabling service objects and active sensors to access management information. This API is made up of five primitives: getAbout(), getVersion(), get(), set() and action(). get() and set() handle read and write objects. The action() method launches code associated with the managed object;

– *Dinfo base manipulation interface:* this API enables access and control of Dinfo objects. It is made up of the following primitives: getAbout(), getVersion(), get(), set(), function(), getObject(), nextObject(), restart() and putObject(). It also makes it possible for Dinfo objects to be manipulated for a specific management activity. For a specific identification of each Dinfo object in the base, a specific identifier is associated with each object. The system used to generate these identifiers is similar to naming trees used by SMI[4] in which leaves represent Dinfo objects. Thus, management applications have the possibility of creating specific information in the active router and to implement them by using these two APIs.

12.3.6. *Definition of QoS policies*

Policies are defined according to approved SLS[5] between the client and ISP. This process is not defined in this book. We consider that this phase has already been completed and that the SLS has already been converted in a group of policy rules. These policy rules are made up of a condition part and an action part. Thus, for example, if the ISP has agreed with one of its users to provide him with a GOLD level service during a specific time slot in a given day, the corresponding policy rule is:

IF (PeriodValidityPolicy = = ValidPeriod) **THEN** Provide User with "GOLD" Service

This business rule is then converted into one of more network level rules. These network rules will make it possible to identify user flows when they enter the network (by corresponding edge router) and to match these flows to the corresponding GOLD service DSCP value. This rule is then created in the CIM

3 *Simple network management protocol.*
4 *System management information.*
5 SLS: *service level specification*, i.e. group of performance parameters contracted by an SLA.

repository and applies to the edge router by which this user's flows access the network. The network rule then generated is:

IF (PeriodValidityPolicy = = ValidPeriod) **and** (IPAddr ∈ {All user IP addresses})
THEN mark Flow with a DSCP = "GOLD"

The basic concepts used here are the same as those introduced by IETF and DMTF policy workgroups. However, in our proposition, it is also possible to use this policy concept for the deployment of new services or active sensors at the level of the managed router. For example, if an ISP has decided, for one of its AF classes, to execute traffic shaping only if more than 80% of the connection's capacity is used, the corresponding policy rule will be defined as follows:

IF (RateLineUsage > Threshold) **THEN** Install(Shaper (TBF(r,b)), AF)

This approach then provides the administrator with a powerful tool to efficiently control the network's behavior, which does not exist in current networks.

12.3.7. *Definition and deployment of TCB*

SCR of DACA architecture described previously contains a set of active codes which implement different DiffServ model algorithms. It provides the ISP with a very flexible approach for defining network behavior. The internal structure of this repository is described in Figure 12.9.

This structure is illustrated in the form of a tree, where the highest level represents the abstract service of the active DiffServ model. Under this entity, we find five different sub-trees, each representing one of the TCB capabilities: *classifier*, *meter*, *action*, *dropper* and *scheduler*. All these entities represent active code containers implementing different types of algorithms.

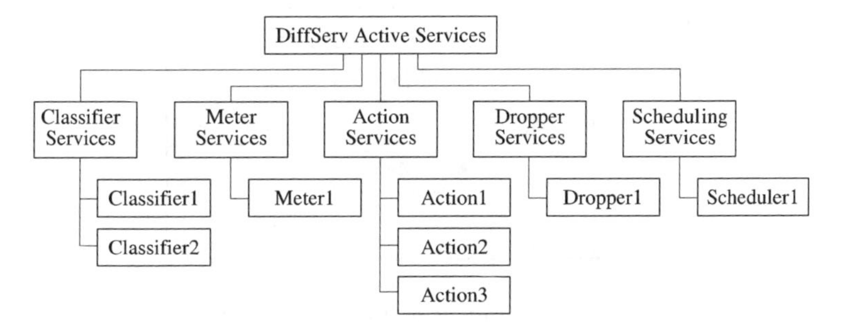

Figure 12.9. *Active TCB services*

TCB in the DiffServ model consists of a group of blocks connected according to a specific sequence depending on the desired behavior. In this case, the administrator defines a set of blocks to deploy in active routers according to these objectives. Once this is done, the administrator delegates TCB installation to the code distribution point of target nodes. This task is achieved by using an active code installation protocol. The selected blocks are then installed in target routers and therefore become visible in the execution environment services block of the active target router. This installation will only affect capsules involved in the installation of this new TCB.

In general, the network administrator defines several TCB classes (called TCBClass). Each class will then be assigned to a specific service and installed in the appropriate active router. A chain of active services will make up each TCBClass.

When the PFM identifies an active capsule (i.e. a capsule belonging to an active protocol present in the node), it sends it to the execution environment. In the opposite case where the capsule is not filtered, it is directly transmitted to the first TCB block. This first block, generally corresponding to the *classifier*, will make it possible, according to the capsule's DSCP value, to apply the appropriate TCB to it. The classifier makes it possible to direct capsules to the appropriate blocks for their processing, i.e. towards the appropriate TCBClass which groups a set of TCBElem (*meter*, *dropper*, *marker*, etc.), as described in Figure 12.10.

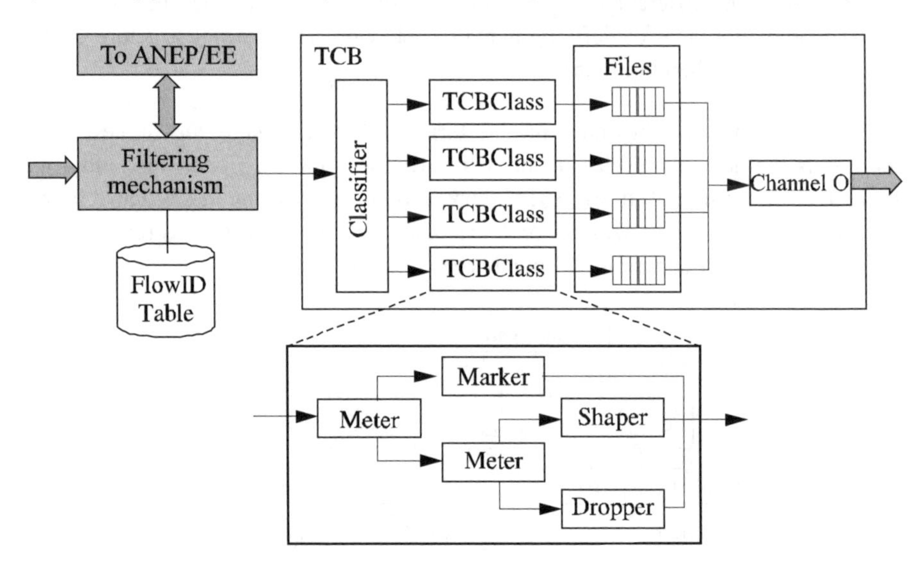

Figure 12.10. *Definition of TCB and block sequence*

If the administrator also wants to automate the installation and configuration process, he will be able to achieve this by defining specific policy rules dedicated to the installation of new TCB. These rules enable the installation, modification or withdrawal of new TCB according to a set of conditions such as network load, performance data retrieved by DMIB, etc. The interaction process between PFM, TCB, execution environment and active applications which enable these tasks to coexist and to be executed are described in Figure 12.11.

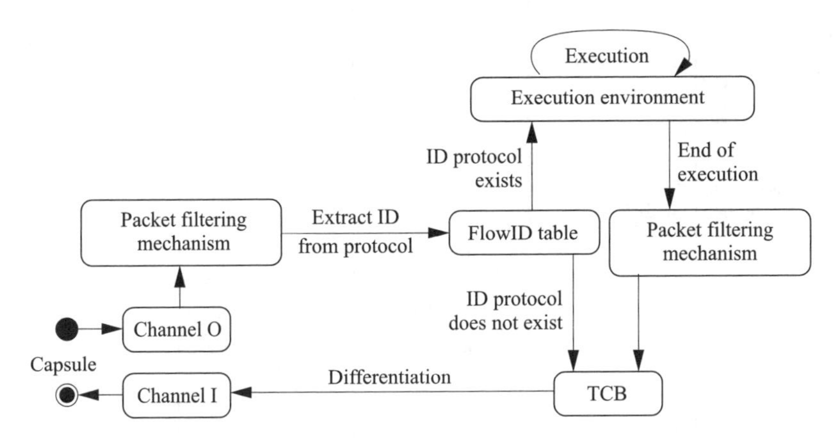

Figure 12.11. *Interaction process between filter, I/O filters, FlowID table, TCB and execution environment*

12.3.8. *Sensor deployment*

When policies are introduced in the CIM repository (CIMR), they are converted into a set of low level policies, as briefly described in section 12.3.6.

When PDP detects the introduction of a new policy, it decides if this policy must be considered locally or if it should be delegated to the LPDP of active routers involved. This decision depends on LPDP capacity to manage it locally, i.e. to have all information necessary for its accomplishment. An example of policies which can be locally processed is a policy changing the measure TCB *meter* algorithm according to a time condition. Policies which involve behavioral variations of the global network cannot be achieved locally and must be managed at the PDP level.

When the LPDP receives a new policy rule to manage, it creates a new active sensor which will handle this policy. According to condition and action parts of the policy rule, the sensor can request the installation of these capabilities.

The LPDP can then interact with the PDP to download locally the missing capabilities from the provider's sensor code repository (SeCR). The diagram for this sequence is illustrated in Figure 12.12.

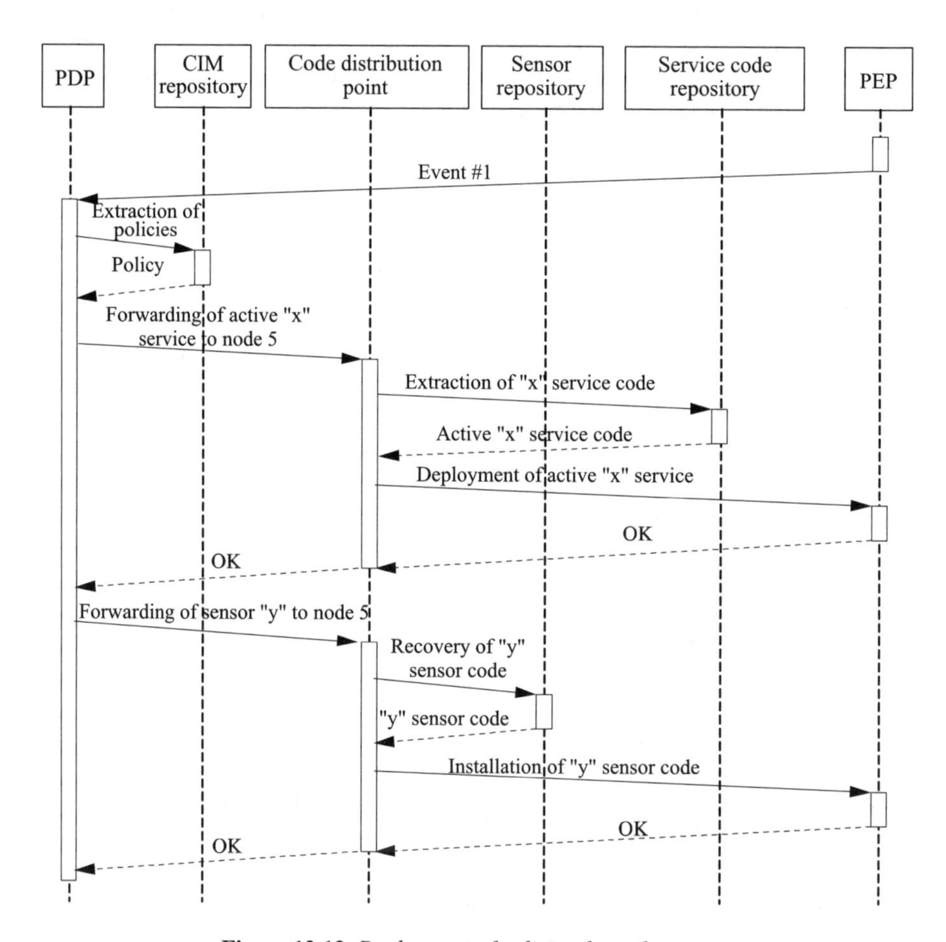

Figure 12.12. *Deployment of policies through sensors*

12.3.9. *Implementation of DACA architecture*

DATA architecture is made up of a group of Java classes developed over the ANTS platform used in the Amarrage project. The principle of the ANTS platform lies in the capability provided to applications to dynamically deploy in the network the protocols that they use through all active routers used by associated flows. ANTS introduces the notion of "*active packet*" *capsule* to replace traditional packets. The capsule is used to carry an application's data between active nodes as well as to transport the code from a protocol to deploy network nodes. These capsules can personalize network services and active nodes. In order to do this, an active node executes each received capsule and maintains a state relative to this execution at active node level. The ANTS platform also uses an efficient and flexible way to deploy codes. When a capsule arrives in a node and requests the execution of an absent protocol, the active node requests the transmission of the associated code from the sending node. This guarantees efficient deployment by recurrence. In our DACA architecture implementation, we have also used the ANEP (*active network encapsulation protocol* [ALE 97]) protocol in order to enable the coexistence of many execution environments (EE) in the same active node. Each ANEP packet uses a specific value in the "*Type ID*" field to identify its own execution environment.

The implementation has consisted of the installation of a group of classes corresponding to different components defined in the architecture represented in Figure 12.2. These classes have been developed in Java2 and extend/ overload or interact with classes native to the ANTS platform. The first class extending the ANTS platform is *activeQoS* class, corresponding to the data plan in DACA architecture. This class is connected to ANTS via 4 Java interfaces: *SendRoEE()*, *ReceiveFromEE()*, *SendToChannel()* and *ReceiveFromChannel()*. We have also extended the capsule class of ANTS by *QoSCapusle* class, containing two additional fields: *FlowIS* and *DCSP*. Fields *FlowIS* and *DCSP* identify the application that generated this capsule and TCB which must be applied to this capsule, respectively. Class *activeQoS* groups several classes. The first, i.e. *Filer*, identifies and then intercepts, with the help of field *FlowID*, capsules belonging to active applications installed in the current DACA node. If the capsule is not intercepted by class *Filter*, then it is transmitted to class *Classifier*, which transmits the capsule to appropriate class *TCBClass* according to *DSCP* field. Each class *TCBClass* is in the form of a *thread* and has a chain of *TCBElems*-type classes, corresponding to different DiffServ algorithms. Each TCBElem applies an "*exec()*" method to the capsule. This method is implemented as an abstract method which must be redefined for each specific DiffServ algorithm (for example, a *meter*, *dropper*, *marker*, etc.). Then, the capsule is inserted in the buffer corresponding to the TCB queue. Finally, we have defined class *SchedulClass* whose goal is to forward capsules to the output channel (*Channel* class in ANTS). As for TCBElem, class SchedulClass possesses an abstract "*Schedul()*" method which must be redefined for each *scheduling* algorithm. For now, DACA architecture implements two different *scheduling* algorithms; the first corresponds to *priority queuing* and the second corresponds to *round robin*. Of course, it is possible to enhance the architecture with new algorithms.

In order to make it possible to execute an on the fly change to algorithms defined in *TCBElem* and *SchedulClass*, the DACA architecture uses a hierarchical repository of DMIB classes in order to manage these classes. This repository is able to register and process services as well as other classes such as PEP services, active probes, MIB, TCBElem, SchedulClass, etc. Any class registered in the DMIB repository must respect a specific interface called "DInfo".

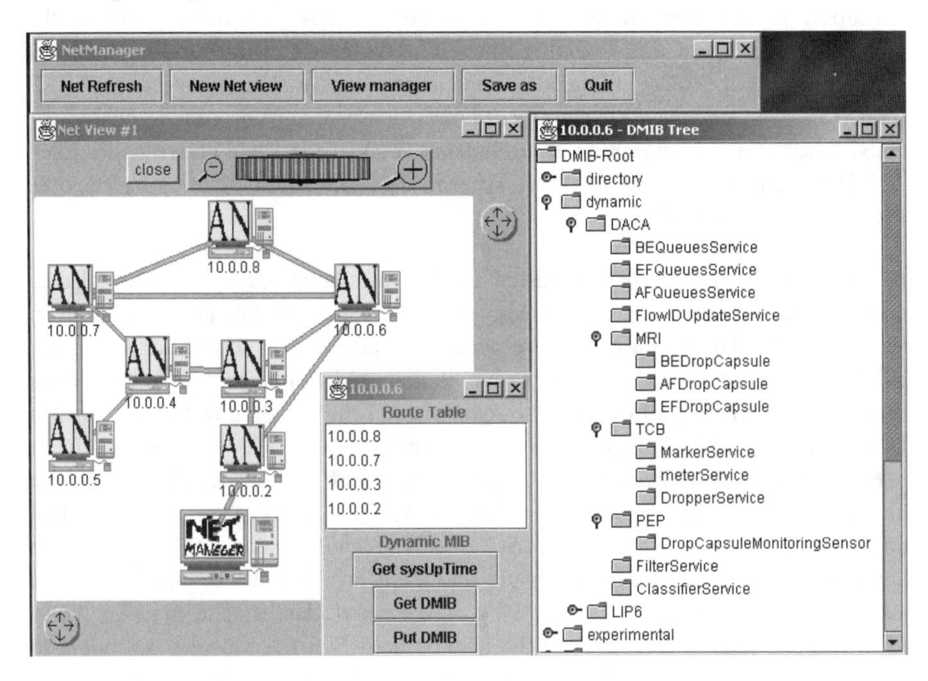

Figure 12.13. *Control interface of active DiffServ services for a network node*

Figure 12.13 gives an overview of the GUI implemented for AMS (*active management system*) of DACA architecture. This interface enables the administrator to install, delete or upgrade active sensors or services in any active network node. In a first window, we find a network topology representation and in a second window a list of different active services installed in the selected active node as well as a group of menus for the manipulation of these remote services.

12.3.10. *Evaluation of DACA architecture behavior*

In order to demonstrate the applicability of our DACA architecture, we have implemented an active DiffServ network benchmark from numerous nodes executing ANTS. The objective is to show the flexibility brought on by DACA system to an operator in the deployment of its operational strategy and the automatic adaptation of the network in relation to a change of strategy.

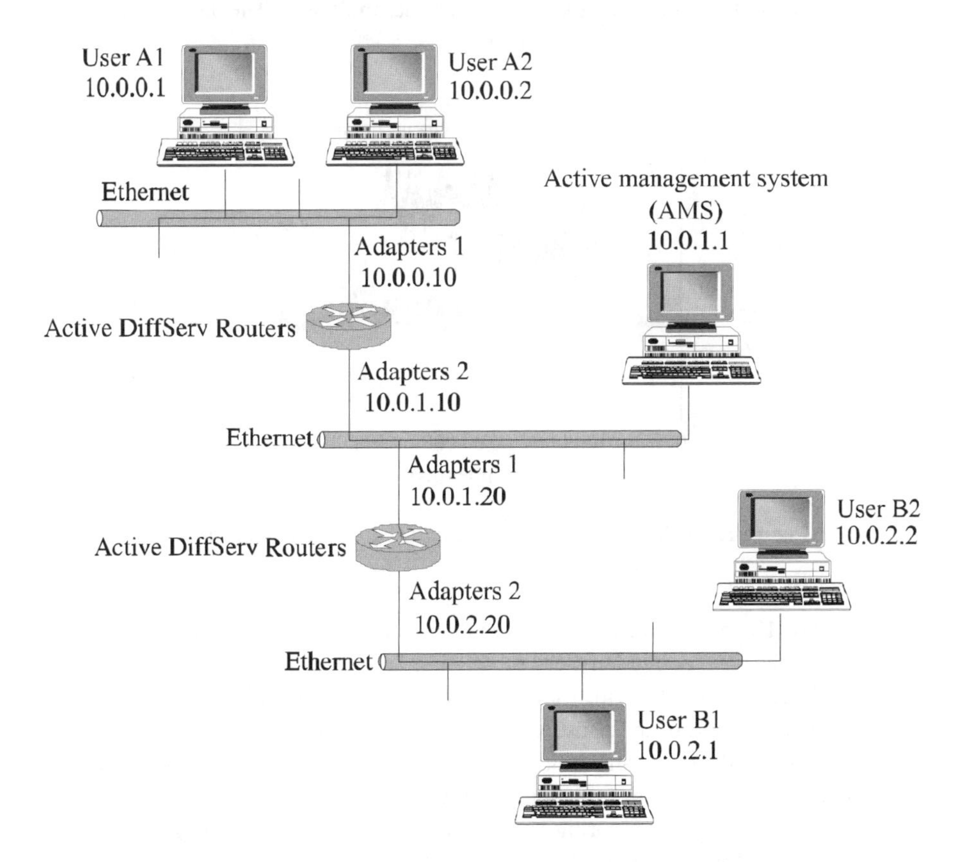

Figure 12.14. *Topology of active router test network*

These tests were completed in the benchmark test presented in Figure 12.14. The equipment used was PCs with Pentium III processors and 10 Mbps Ethernet adapters. In this scenario, subscribers A1 and B1 respectively launch a communication session with subscribers A2 and B2. In this scenario, we presume that the operator implements at first a simple strategy consisting of no differentiation between services provided to its users. Then, the strategy changes and the operator wants to implement a differentiation of services. The performance measured is that of round trip time (RTT) between both user stations when the administrator applies different strategies.

When the administrator implements a strategy of non-differentiation, the scheduling algorithm used is *round robin* but when the strategy changes, the scheduling algorithm uses the algorithm by priority. The implementation of the new strategy does not require any manual router manipulation. The heterogenity and distribution aspects are completely hidden.

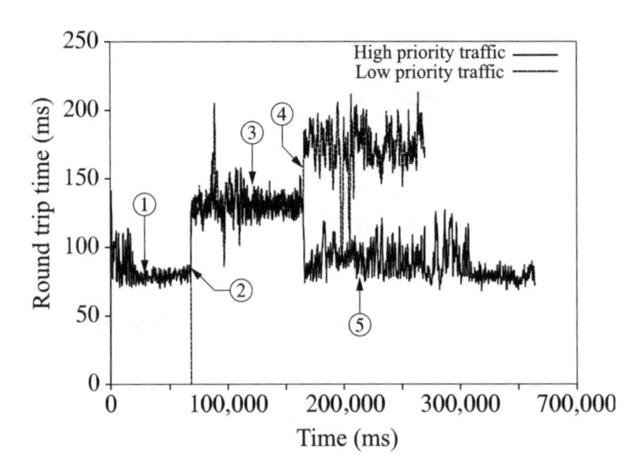

Figure 12.15. *Example of deployment of a new differentiation strategy*

In Figure 12.15, we can identify three different network behaviors with DACA. In the first phase (1), active DiffServ routers use an efficient round robin scheduling service which only works with traffic A1-A2. In this case, RTT for this traffic is approximately 80 ms. In the phase starting at point (2), we notice that launching of traffic B1-B2 will create an additional delay (60 ms) for flow A1-A2 and will deteriorate delays (3). From this moment, the administrator decides to implement a differentiation strategy by favoring traffic from *gold* subscribers willing to pay more for a better quality of service (subscribers A). Thus, the administrator uses the DACA system to deploy on the fly a new algorithm which enables the differentiation between high priority flows and low priority flows according to a specific packet marking for each subscriber. The immediate result obtained shows that from this moment (4), high priority traffic is controlled differently and obtains more resources than low priority traffic (5). High priority traffic RTT decreases until it reaches a value close to 80 ms to the detriment of low priority traffic RTT which reaches a value close to 150 ms. This result validates on the one hand good high priority scheduling algorithm behavior and, on the other hand, flexibility of deployment of a new strategy by the administrator because of DACA. Other similar experiments have been accomplished to validate the functioning of DACA architecture in its different perspectives, particularly the flexibility it offers in controlling the behavior of DiffServ networks.

12.4. Conclusion

In this chapter, we have presented contributions from the French community in active network field. More specifically, we have focused on the first active network project (the Amarrage project) from the RNRT program supported by the research ministry in France. The goal of this project has been to explore the potential of active networks for various transport, QoS, video and network management problems.

We have then presented one of the results from the Amarrage project: the DACA architecture. The goal of the DACA architectural solution is to associate the approach of policy-based management with active network technology in order to dynamically control DiffServ networks. It consists of only using active technology in a control plan to maintain compatibility with the current Internet, while enabling a smooth migration to all active. This architecture uses control and management applications as well as active protocols to deploy and configure the network. DiffServ mechanisms are defined as active services whose life cycle can be managed by the administrator according to needs. Policies enable the administrator to automate deployment and configuration processes in a very flexible manner. In this architecture, active network technology provides the level of flexibility required and its absence would greatly penalize traditional approaches. It provides an efficient solution for monitoring and optimization of the use of resources via a simplified control by the use of policy rules. This architecture has been implemented in the ANTS environment and validation tests of this concept have been completed.

12.5. Bibliography

[ALE 97] ALEXANDER D.S., *et al.*, "Active Network Encapsulation Protocol", *Draft – DARPA AN Working Group*, July 1997.

[BLA 98] BLAKE S., BLACK D., CARLSON M., DAVIES E., WANG Z., WEISS W., "An Architecture for Differentiated Services", *RFC 2475*, December 1998.

[CHA 01] CHAN K., *et al.*, "COPS Usage for Policy Provisioning (COPS-PR)", *RFC 3084*, March 2001.

[DUR 00] DURHAM D., *et al.*, "The COPS (Common Open Policy Service) Protocol", *RFC 2748*, January 2000.

[HER 00] HERZOG S., *et al.*, "COPS usage for RSVP", *RFC 2749*, January 2000.

[MPE 98] MPEG-4, "Coding of audio visual objects: Visual, Technical report 14496-2 (MPEG-4) – ISO/IEC JTC1/SC29/WG11 N2202", Tokyo, Japan, March 1998.

[NIC 99] NICHOLS K., JACOBSON V., ZHANG L., "A Two-bit Differentiated Services Architecture for the Internet", *RFC 2638*, July 1999.

[PRO 00] PROJET AMARRAGE, "Architecture multimédia & administration réparties sur un réseau actif à grande échelle", *Pre-competitive project no. 41 RNRT research program*, Paris, France, 2000-2002, URL: http://www.telecom.gouv.fr/rnrt/projets/res_d41_ap99.htm.

[RAJ 99] RAJAN R., VERMA D., KAMAT S., FELSTAINE E., HERZOG S., "A policy framework for integrated and differentiated services in the Internet", *IEEE Network Magazine 13*, no. 5, p. 36-41, September/October 1999.

[REJ 96] RIJKSE K., "H.263: video coding for low-bit-rate communication", *IEEE Comm.*, vol. 34, no. 12, p. 42-45, December 1996.

[WET 98] WETHERALL D., GUTTAG J., TENNENHOUSE D., "ANTS: A Toolkit for Building and Dynamically Deploying Network Protocols", *IEEE OPENARCH'98*, San Francisco, USA, April 1998.

List of Authors

Nadjib ACHIR
Institut Galilée
University of Paris 13
France

Badr BENMAMMAR
Labri
ENSEIRB
Bordeaux
France

Miguel CASTRO
INT
Evry
France

Zeina EL FERKH JRAD
LIPN
University of Paris 13
France

Mauro FONSECA
CEFET-PR
Curitiba
Brazil

Dominique GAÏTI
University of Technology of Troyes
France

Yacine GHAMRI-DOUDANE
IIE
Evry
France

Francine KRIEF
Labri
ENSEIRB
France

Hugues LECARPENTIER
LIP6
Pierre and Marie Curie University
Paris
France

Anneli LENICA
France Telecom
Issy les Moulineaux
France

Leila MERGHEM-BOULAHIA
University of Technology of Troyes
France

Nada MESKAOUI
University of Beirut
Lebanon

Abdallah M'HAMED
INT
Evry
France

Hassine MOUNGLA
LIPN
University of Paris 13
France

Sidi-Mohammed SENOUCI
France Telecom
Lannion

Djamal ZEGHLACHE
INT
Evry
France

Index

E, F

H, I

K, L, M

N, O

P

Q

R

S

T, U

V, W